Hydraulic Design Series No. 4

Introduction to Highway Hydraulics

1. Report No. FHWA-NHI-01-019	2. Government Accession No.	3. Recipient's Catalog No.	
4. Title and Subtitle Introduction to Highway Hydraulics Hydraulic Design Series Number 4 (HDS 4)		5. Report Date August 2001	
		6. Performing Organization Code	
7. Author(s) James D. Schall, Everitt V. Richardson, and Johnny L. Morris		8. Performing Organization Report No.	
9. Performing Organization Name and Address Ayres Associates Fort Collins, Colorado 80527		10. Work Unit No. (TRAIS)	
		11. Contract or Grant No.	
12. Sponsoring Agency Name and Address National Highway Institute (NHI) and Office of Bridge Technology (HIBT) Federal Highway Administration Washington, D.C. 20590		13. Type of Report and Period Covered	
		14. Sponsoring Agency Code	
15. Supplementary Notes FHWA COTR: Jorge Pagán-Ortiz, FHWA, Washington, D.C. Technical Assistance: Philip Thompson, FHWA, Washington, D.C.			

16. Abstract

Hydraulic Design Series No. 4 provides an introduction to highway hydraulics. Hydrologic techniques presented concentrate on methods suitable to small areas, since many components of highway drainage (culverts, storm drains, ditches, etc) service primarily small areas. A brief review of fundamental hydraulic concepts is provided, including continuity, energy, momentum, hydrostatics, weir flow and orifice flow. The document then presents open channel flow principles and design applications, followed by a parallel discussion of closed conduit principles and design applications. Open channel applications include discussion of stable channel design and pavement drainage. Closed conduit applications include culvert and storm drain design. Examples are provided to help illustrate important concepts. An overview of energy dissipators is provided and the document concludes with a brief discussion of construction, maintenance and economic issues.

As the title suggests, Hydraulic Design Series No. 4 provides only an introduction to the design of highway drainage facilities and should be particularly useful for designers and engineers without extensive drainage training or experience. More detailed information on each topic discussed is provided by other Hydraulic Design Series and Hydraulic Engineering Circulars.

This publication is an update of the 1997 edition, and now includes dual units (SI and English). HDS-4 was first published in 1965 under the name Design of Roadside Drainage Channels," which concentrated on open channel hydraulics and did not include discussion of closed conduit facilities and energy dissipators.

17. Key Words Hydrology, hydraulics, highway drainage, open channels, roadside ditches, pavement drainage, inlets, closed conduits, culverts, storm drains, energy dissipators		18. Distribution Statement This document is available to the public from the National Technical information service, Springfield, Virginia 22161	
19. Security Classif. (of this report) Unclassified	20. Security Classif. (of this page) Unclassifed	21. No. of Pages 214	22. Price

Form DOT F 1700.7 (8-72)   Reproduction of completed page authorized

TABLE OF CONTENTS

LIST OF FIGURES

vi

LIST OF TABLES

(page intentionally left blank)

LIST OF SYMBOLS

A   =   Cross-sectional area, or drainage area

B   =   Bottom width of a channel

C   =   Runoff coefficient in Rational Method

D   =   Culvert diameter or sediment size

D_{50} =   Sediment size for which 50% by weight of the particles are are smaller; similarly D_{65}, D_{84}, D_{90} represent sizes for which 65, 84, and 90% of the particles are smaller

Fr   =   Froude Number

f   =   Darcy-Weisbach friction factor

g   =   Acceleration due to gravity, 9.81 m/s^2 (32.2 ft/s^2)

H   =   Specific energy

H_f =   Friction loss of head

h_L =   Headloss

i   =   Rainfall intensity, mm/h

K   =   Conveyance

k_s =   Height of roughness channels

K_u =   Correction factor for units: SI (English)

n   =   Manning's resistance coefficient

P   =   Wetted perimeter of the flow

Q   =   Discharge (flow rate)

q   =   Unit discharge

R   =   Hydraulic radius

Re   =   Reynolds Number

r_c =   Radius of curvature at the center of the stream

r_i, r_o =   Radius of curvature for inner and outer banks in a bend

S   =   Channel slope

S_f	=	Friction slope (also called the energy slope)
S_o	=	Bed slope
S_w	=	Slope of water surface
T	=	Top width
V	=	Velocity
V_*	=	Shear velocity, $\sqrt{\tau_o/\rho}$
X	=	Coordinate axis
y	=	Coordinate axis, depth of flow
y_c	=	Critical depth of flow
y_s	=	Scour depth
Z	=	Coordinate axis, side slope of channel bank
z	=	Elevation above arbitrary reference level

GREEK SYMBOLS

α	=	Energy correction factor
β	=	Momentum correction factor
γ	=	Specific weight of water, 9810 N/m^3 (62.4 lbs/ft^3)
λ	=	Wave length
μ	=	Dynamic viscosity of a fluid
ν	=	Kinematic viscosity of a fluid
ρ	=	Fluid density, water = 1,000 kg/m^3 (1.94 slugs/ft^3)
ρ_s	=	Density of sediment particles
τ	=	Shear stress
τ_o	=	Bed shear stress
τ_c	=	Critical shear stress of sediment particles

ACKNOWLEDGEMENTS

This manual is a dual-unit update of the June 1997 version (FHWA Publication FHWA HI 97-028). The authors wish to acknowledge the technical assistance on the 1997 document provided by Abbi Ginsberg, FHWA, and Chris Dunn and Arlo Waddoups, formerly FHWA. The 1997 document was a major update and expansion of the original HDS-4, that was written by James Searcy, FHWA, and published in 1965 under the title "Design of Roadside Drainage Channels (Government Printing Office Stock Number 050-001-00068-7, Catalog Number TD 2.33:4).

(page intentionally left blank)

1. INTRODUCTION

1.1. General

Highway hydraulic structures perform the vital function of conveying, diverting, or removing surface water from the highway right-of-way. They should be designed to be commensurate with risk, construction cost, importance of the road, economy of maintenance, and legal requirements. One type of drainage facility will rarely provide the most satisfactory drainage for all sections of a highway. Therefore, the designer should know and understand how different drainage facilities can be integrated to provide complete drainage control.

Drainage design covers many disciplines, of which two are hydrology and hydraulics. The determination of the quantity and frequency of runoff, surface and groundwater, is a hydrologic problem. The design of structures with the proper capacity to divert water from the roadway, remove water from the roadway, and pass collected water under the roadway is a hydraulic problem.

This publication will briefly discuss hydrologic techniques with an emphasis on methods suitable to small drainage areas, since many components of highway drainage (e.g., storm drains, roadside ditches, etc.) service primarily small drainage areas. Fundamental hydraulic concepts are also briefly discussed, followed by open-channel flow principles and design applications of open-channel flow in highway drainage. Then, a parallel discussion of closed-conduit concepts and applications in highway drainage will be presented. The concluding sections include an introduction to energy dissipation, construction, maintenance, and economic issues. In all cases, detailed design criteria and standards are provided primarily by reference, since the objective of this document is to present a broad overview of all the components of highway drainage, and to serve primarily as an "Introduction to Highway Hydraulics."

1.2. Types of Drainage Facilities

Highway drainage facilities can be broadly classified into two major categories based on construction: (1) open-channel or (2) closed-conduit facilities. Open-channel facilities include roadway channels, median swales, curb and gutter flow, and others. Closed-conduit facilities include culverts and storm drain systems. Note that from a hydraulic classification of flow condition, open-channel or free-surface flow can occur in closed-conduit facilities.

Figure 1 shows a typical divided highway where a variety of open-channel and closed-conduit facilities are needed to drain the highway. Starting at the outer edge of the right-of-way are the intercepting channels on the natural ground outside the cut-and-fill or on benches breaking the cut slope. In an arid region, intercepting channels (or dikes) may also be used for great distances along the roadway to capture overland flow runoff from large upstream watersheds. Next are the roadway channels between the cut slope and shoulder of the road and the toe-of-slope channels which take the discharges from the roadway channels and convey it along or near the edge of the roadway embankment to a point of disposal. A shallow depression or swale drains the median to an inlet that conveys water to the culvert. The culvert itself provides for cross drainage of a relatively large stream channel.

1

Figure 1. Types of highway drainage facilities.

1.3. Design Philosophy

The primary purpose of highway drainage facilities is to prevent surface runoff from reaching the roadway, and to efficiently remove rainfall or surface water from the roadway. Designing drainage facilities to accomplish this purpose requires balancing the risk of future damages from runoff events (whose occurrence, in time or magnitude, cannot be forecast accurately) against the initial construction cost. Since this is not easy to accomplish, it is customary that a particular flood frequency be selected for each class of highway to establish the design discharge for sizing drainage facilities. This design frequency is then adjusted based on evaluating a check flood to better account for the risk involved considering traffic conditions, structure size, and value of adjacent property.

For costly or high risk facilities, a range of discharges with a range of flood frequencies are used to evaluate drainage facilities. The range of floods considered usually includes the "base flood" and sometimes the "super flood." The base flood is defined as the flood (storm or tide) having a one percent chance of being equaled or exceeded in any given year. This flood is also referred to as the 100-year flood, meaning that over an infinite period of time this flood will be equaled or exceeded on average once every 100 years. A super flood is significantly greater than the base flood. For example, a 0.2 percent discharge (a 500-year flood) is one possible super flood. Note that it is seldom possible to compute a 0.2 percent discharge with the same accuracy as the one percent flood; nevertheless, it does draw attention to the fact that floods greater than the one percent flood can occur. Such floods (super floods) may be defined as a flood exceeding the base flood, whose magnitude is subject to the limitation of state-of-the-art analytical practices. Other floods considered include the overtopping flood, maximum historical flood, probable maximum flood, and design flood.

A range of floods is also typically assumed to evaluate pavement drainage design. Pavement drainage is normally designed for the 10-year flood, except in sag vertical curves where water cannot escape other than through a storm drain. In these locations, a 50-year event is often used for design to prevent ponding to a depth that could drown people if they were to

2

drive into it. The use of a lesser frequency event, such as the 50-year storm, to assess hazards at critical locations where water can pond to appreciable depths is commonly referred to as the check storm or check event. The spread of water on the pavement during a check storm can also be evaluated, with a typical criteria being at least one lane of traffic open during the check event.

One way to select the design flood frequency is through the concept of economics by establishing the least total expected cost for the structure. This concept considers the capital costs, maintenance costs, and the flood hazard costs that are incurred due to damage by a range of flooding events. The flood frequency that generates the least total expected cost for the life of the project would be the one chosen for the design of the structure.

1.4. Metric System

The metric system of measurement has been in use for hundreds of years. The SI metric system is an abbreviation for International System of Units, a modern day metric system established by international agreement in 1960. The SI system has been favorably received throughout the world and provides a standard international language to describe measurement.

Appendix A summarizes the use of SI, particularly as it relates to highway drainage design. This document will be presented in dual units with SI as the primary unit. The English system of units, which will follow in parentheses, refers to the U.S. customary units.

(page intentionally left blank)

2.   ESTIMATING STORM RUNOFF FROM SMALL AREAS

2.1.   General

The first step in designing a drainage facility is to determine the quantity of water the facility must carry. The hydrologic analysis required to estimate discharge can be a major component of the overall design effort. The level of effort required depends on the available data and the sophistication of the analytical technique selected. Regardless of the analytical technique used, hydrologic analysis always involves engineering judgment due to the complexity and inherent random nature of the runoff process itself. Unlike many other aspects of engineering design, the quantification of runoff is not a precise science.

For routine design problems, particularly involving small drainage areas, it is impractical and unnecessary to use sophisticated analytical methods that require extensive time and labor. Fortunately, there are a number of sound and proven methods available to analyze hydrology for the more traditional and routine day-to-day design problems. These procedures enable peak flows and hydrographs, a plot of the variation of discharge with time (figure 2), to be determined without an excessive expenditure of time. They use existing data, or in the absence of data, synthetic methods to develop design parameters.

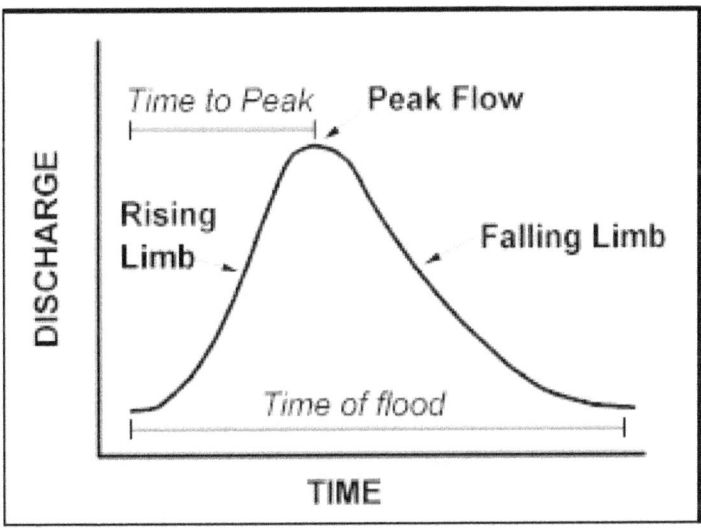

Figure 2. Flood hydrograph.

Drainage design for facilities serving small areas can typically be based on peak flow conditions. Knowledge of the complete hydrograph is seldom necessary for small drainage facilities. For example, the design of median drainage facilities, a storm drain and inlets to protect a fill slope, or a culvert draining a small area isolated by roadway fill, can all be designed based on peak flow conditions only. Information in this chapter summarizes standard methods for estimating peak flow. A more comprehensive treatment of peak flow estimation, and a complete discussion of hydrograph calculation and other hydrologic topics, is provided in Hydraulic Design Series Number 2 entitled, Highway Hydrology (HDS-2).[1]

Methods for making peak flow estimates can be separated into two categories: (1) sites with measured stream gage data, and (2) sites without gaged data. When gaged data of adequate length and quality are available, statistical analysis of the flow record can be used to estimate flood peaks for various return periods. Generally speaking, gaged data are available only for larger stream channels; consequently, there are only limited opportunities to apply this method in smaller watersheds, and practically no application of this method to the small areas which contribute runoff to highway drainage structures. Techniques for analysis of gaged data are briefly summarized below and a comprehensive treatment is provided in HDS-2.[1]

When gaged data are not available, estimates are made by empirical equations (e.g., Rational Method) or by regional regression equations. Regional regression equations are typically appropriate for larger drainage areas, and methods such as the Rational Method are commonly used for smaller areas, less than about 80 hectares or 200 acres. For larger ungaged areas, regression equations developed from regional data are recommended. Both procedures are outlined in this chapter. Note that there is no clearly defined line where one method should end and another method be used. The methods sometimes give results for the same area that agree quite well, and in other instances, they may disagree by 50 or more percent. When major differences occur, the applicability of each method should be evaluated and significant engineering judgment will be required to establish reasonable design values.

The recurrence interval of the design discharge (see section 1.2) is of concern because economy is always a factor in the design, as discussed in chapter 11. The recurrence interval defines the frequency that a given event (e.g., rainfall or runoff) is equaled or exceeded on the average, once in a period of years. For example, if the 25-year frequency discharge is 100 m^3/s (3,531 ft^3/s), a runoff event of this size or greater would be expected to occur on average once every 25 years. The exceedence probability, which is the reciprocal of recurrence interval, is also used in design. For the above example, a discharge equal to or greater than 100 m^3/s (3,531 ft^3/s) would have a 0.04 probability, or a 4 percent chance of occurrence in any given year.

Overdesign and underdesign both involve excessive costs on a long-term basis. A channel designed to carry a 1-year flood would have a low first cost, but the maintenance cost would be high because the channel and roadway may be damaged by storm runoff almost every year. On the other hand, a channel designed to carry the 100-year flood would be high in first cost, but low in maintenance cost. Somewhere between these limits lies the design frequency which will produce a reasonable balance of construction cost, annual maintenance cost, and risk of flooding.

2.2   Storm Runoff

Precipitation falling on land and water surfaces produces watershed runoff. A small part of the precipitation evaporates as it falls and some is intercepted by vegetation. Of the precipitation that reaches the ground, a portion infiltrates the ground, a portion fills the depressions in the ground surface, and the remainder flows over the surface (overland flow) to reach defined watercourses. The surface runoff is sometimes augmented by subsurface flow that flows just beneath the ground surface and reaches the watercourse in time to be a part of the storm runoff.

The precipitation infiltrating the ground replenishes the soil moisture and adds to the groundwater storage. Some of this underground water reaches the stream long after the storm runoff has passed and some is withdrawn by the life processes of vegetation or by man for his use.

The storm runoff which must be carried by highway drainage facilities is thus the residual of the precipitation after losses (the extractions for interception, infiltration, and depression storage). The rate of water loss depends upon the amount of the precipitation and the rate at which it falls (intensity), upon temperature, and the characteristics of the land surface. Not only does the rate of runoff vary with the permeability of the land surface and the vegetal cover, but it varies with time for the same surface depending upon the antecedent conditions, such as soil moisture, etc.

2.3. Analysis of Gaged Data

The U.S. Geological Survey (USGS) collects and publishes much of the stream gage data available in the United States. These data are reported in USGS Water Supply Papers (by state), in Annual Surface Water Records and on computer files. Statistical analysis of gaged data permits an estimate of the peak discharge in terms of its probability or frequency of occurrence at a given site.

The frequency distributions that have been found most useful in hydrologic data analysis are the normal distribution, the log-normal distribution, the Gumbel extreme value distribution and the log-Pearson Type III distribution. The log-Pearson III has been widely used for flood analyses and the U.S. Water Resources Council has recommended it as the standard distribution for flood frequency analyses. A comprehensive treatment on the use of this distribution in the determination of flood frequency distributions is presented in Interagency Advisory Committee on Water Data, Hydrology Committee Bulletin 17B.[2] A complete treatment of the statistical analysis of gage data is provided by HDS-2.[1]

2.4. Rainfall Intensity-Duration-Frequency Analysis

The intensity of rainfall is the rate at which rain falls. Intensity is usually stated in mm/h (in/hr) regardless of the duration of the rainfall, although it may be stated as total rainfall in a particular time period (i.e., duration). Frequency can be expressed as the probability of a given intensity of rainfall being equaled or exceeded, or it can be expressed in terms of the average interval (recurrence interval) between rainfall intensities of a given or greater amount. The frequency of rainfall intensity cannot be stated without specifying the duration of the rainfall because the rainfall intensity varies with the duration of rainfall (figure 3).

Point rainfall data are used to derive intensity-duration-frequency curves necessary in hydrologic analysis (e.g., as required in the Rational Method; see section 2.5). Two methods for selecting rainfall data used in frequency analyses are the annual series and the partial-duration series. The annual-series analysis considers only the maximum rainfall for each year (usually calendar year) and ignores the other rainfalls during the year. These lesser rainfalls during the year sometimes exceed the maximum rainfalls of other years. The partial-duration series analysis considers all of the high rainfalls regardless of the number occurring within a particular year. In designing highway drainage facilities for return periods greater than 10 years, the difference between the two series is unimportant. When the return period (design frequency) is less than 10 years, the partial-duration series is believed to be more appropriate. To change the frequency curves based on the annual series to one based on the partial-duration series, multiply the annual series values by the following factors:[3]

2-year return period	1.14
5-year return period	1.04
10-year return period	1.01
20-year or more	1.00

7

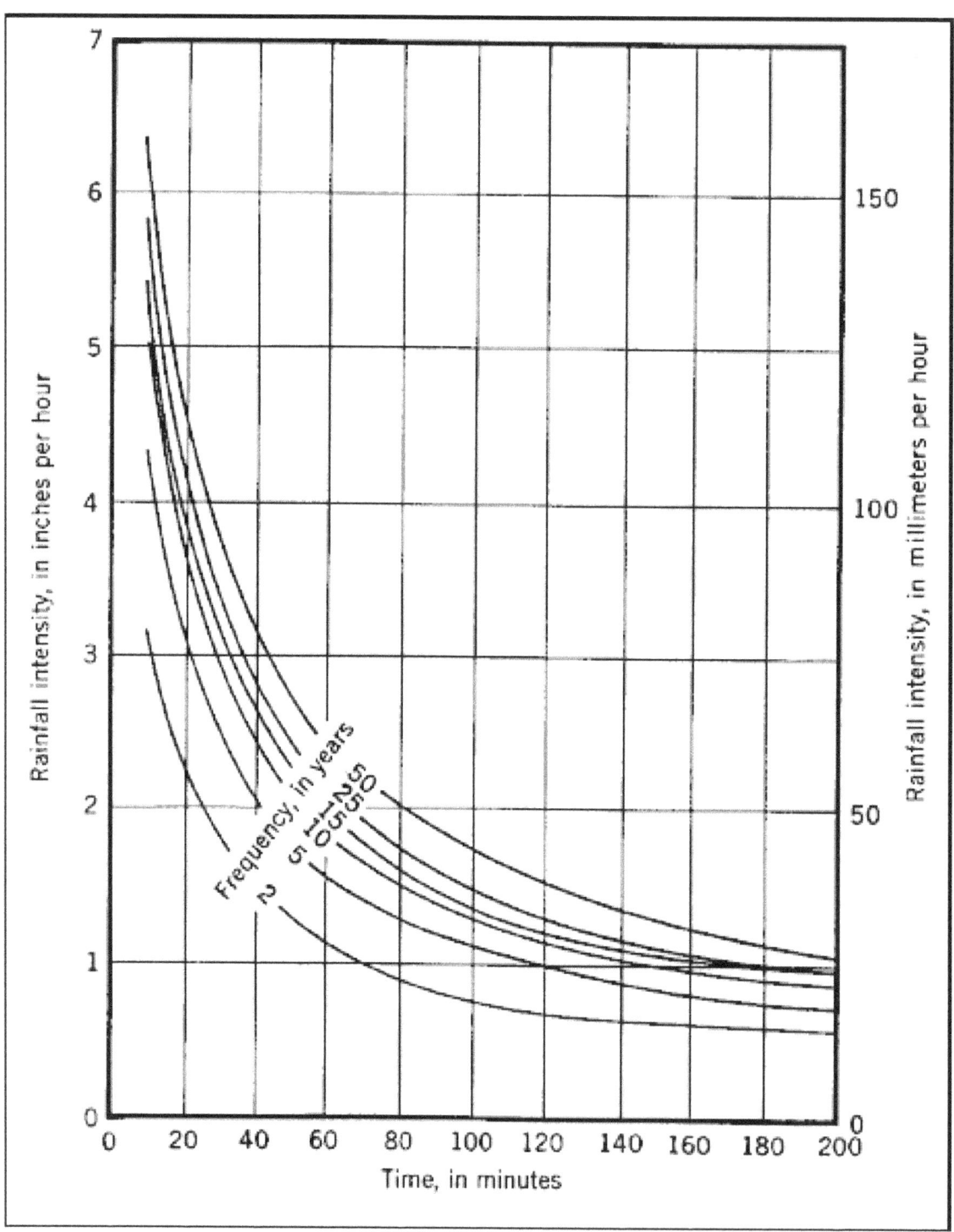

Figure 3. Typical Intensity-duration rainfall curves.

8

Point rainfall data are collected at approximately 20,000 locations every day by the National Weather Service (NWS), the National Oceanic and Atmospheric Administration (NOAA) and other agencies. The data are sent to the Environmental Data and Information Service (EDIS), which has responsibility for processing and disseminating environmental data and is an excellent source of basic rainfall data for highway drainage design. However, in most localities the necessary rainfall information, such as intensity-duration-frequency curves, are available from a city, county or state agency and it is seldom necessary to begin hydrologic analysis with raw rainfall data. If the necessary data are not available locally, a variety of publications are available with rainfall data as summarized in the AASHTO "Highway Drainage Guidelines."[4]

2.5. The Rational Method

2.5.1. Equation and Assumptions

One of the most common equations for peak flow estimation is the rational formula:

$$Q = \frac{CiA}{K_u}$$

(1)

where:

Q   =   Peak rate of runoff, m^3/s (ft^3/s)

C   =   Dimensionless runoff coefficient assumed to be a function of the cover of the watershed

i   =   Average rainfall intensity, for the selected frequency and for duration equal to the time of concentration, mm/h (in/hr)

A   =   Drainage area, tributary to the point under design, hectares (acres)

K_u =   360 (1)

The rational formula assumes that if a uniform rainfall of intensity (i) were falling on an area of size (A), the maximum rate of runoff at the outlet to the drainage area would be reached when all portions of the drainage area were contributing; the runoff rate would then become constant. The time required for runoff from the most hydraulically remote point (point from which the time of flow is greatest) of the drainage area to arrive at the outlet is called the time of concentration (t_c).

Actual runoff is far more complicated than the rational formula indicates. Rainfall intensity is seldom the same over an area of appreciable size or for any substantial length of time during the same storm. Even if a uniform intensity of rainfall of duration equal to the time of concentration were to occur on all parts of the drainage area, the rate of runoff would vary in different parts of the area because of differences in the characteristics of the land surface and the nonuniformity of antecedent conditions.

Under some conditions maximum rate of runoff occurs before all of the drainage area is contributing (see section 2.5.6). The temporary storage of stormwater enroute toward defined channels and within the channels themselves, accounts for a considerable reduction in the peak rate of flow except on very small areas. The error in the runoff estimate increases as the size of the drainage area increases. For these reasons, the rational method should not be used to determine the rate of runoff from large drainage areas. For the design of highway drainage structures, the use of the rational method should be restricted to drainage areas less than 80 hectares (200 acres).

In summary, the assumptions involved in using the Rational Method are:

1.   The peak flow occurs when all the watershed is contributing.

2.   The rainfall intensity is uniform over a time duration equal to the time of concentration, which is the time required for water to travel from the most hydraulically remote point to the outlet or point of interest. Note that the most hydraulically remote point is defined in terms of time, not necessarily distance.

3.   The frequency of the computed peak flow is equal to the frequency of the rainfall intensity. In other words, the 10-year rainfall intensity is assumed to produce the 10-year flood.

2.5.2. Runoff Coefficient

The runoff coefficient (C) in the rational formula is the ratio of the rate of runoff to the rate of rainfall at an average intensity (i) when all the drainage area is contributing. The runoff coefficient is tabulated as a function of land use conditions; however, the coefficient is also a function of slope, intensity of rainfall, infiltration and other abstractions. The range in values of C listed in table 11 in appendix B permit some allowance for land slope and differences in permeability for the same type cover. For flat slopes and permeable soil, use the lower values.

Where the drainage area is composed of several land use types, the runoff coefficient can be weighted according to the area of each type of cover present (see example 2.1). However, the accuracy of the Rational Method is better when the land-use is fairly consistent over the entire area. Significantly different land use conditions can lead to inconsistent estimates of the time of concentration, and hence the intensity, and errors in establishing the most appropriate C.

2.5.3. Time of Concentration

The time of concentration (defined in section 2.5.1) varies with the size and shape of the drainage area, the land slope, the type of surface, the intensity of rainfall, and whether flow is overland or channelized. The time of concentration can be considered the sum of an overland flow time and the travel times in gutters, swales, storm drains, etc. Technically, the time of concentration is the travel time of a wave, as opposed to the actual water velocity, from the most hydraulically remote point to the point in question. However, uncertainties about the actual overland flow path, roughness, slope and rainfall variations (temporal and spatial) limit both the need for making this distinction and the accuracy of any calculation.[5] Extreme precision is not warranted in determining time of concentration, particularly for small area drainage facility design; however, since the peak discharge (Q) is generally quite sensitive to the time of concentration, care should be taken to insure an appropriate value is obtained.

EXAMPLE PROBLEM 2.1 (SI Units)

Given: A toe-of-slope channel collects runoff from the roadway and an adjacent watershed. The tributary area has a fairly uniform cross section as follows: 3.5 m of concrete pavement; 8 m gravel shoulder, channel, and backslope; 60 m of forested watershed. The length of the area is 125 m.

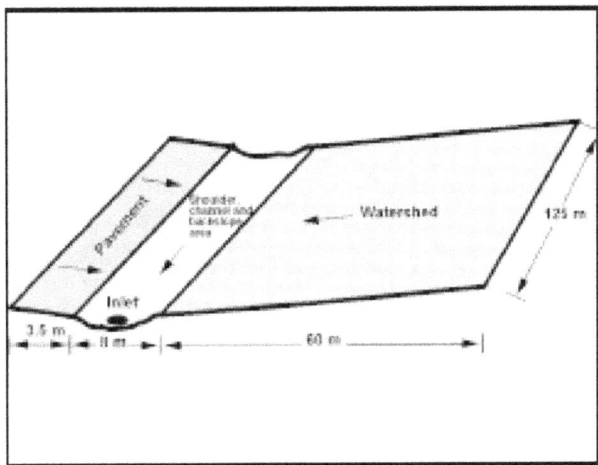

Find: Runoff coefficient, C

Solution:

Type of Surface	(Table B.1)	C Area (hectare)	CA (hectare)
Concrete pavement	0.9	0.043	0.039
Shoulder, channel, and backslope	0.5	0.100	0.050
Forested watershed	0.3	0.750	0.225
TOTAL	---	0.90	0.314

Weighted $C = \dfrac{.314}{0.90} = 0.35$

11

EXAMPLE PROBLEM 2.1 (English Units)

Given: A toe-of-slope channel collects runoff from the roadway and an adjacent watershed. The tributary area has a fairly uniform cross section as follows: 12 ft of concrete pavement; 26 ft gravel shoulder, channel, and backslope; 200 ft of forested watershed. The length of the area is 400 ft.

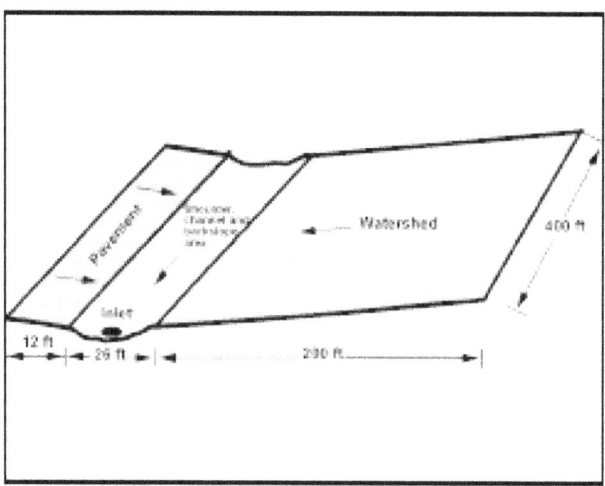

Find: Runoff coefficient, C

Solution:

Type of Surface	C (Table B.1)	Area (acres)	CA (acres)
Concrete pavement	0.9	0.11	0.10
Shoulder, channel, and backslope	0.5	0.24	0.12
Forested watershed	0.3	1.84	0.55
TOTAL	---	2.19	0.77

Weighted $C = \dfrac{0.77}{2.19} = 0.35$

A number of empirical formulas have been proposed for estimating the time of concentration. When the drainage area consists of different flow paths, the time of concentration is the sum of the incremental travel times computed for each different reach. The travel time in gutter, storm drain and channel flow is typically estimated from basic hydraulic data (t = distance/velocity). The velocity in shallow concentrated flow may be estimated from figure 4. For overland flow the most physically correct approach is based on kinematic wave theory:[5]

$$t = K_u \frac{n^{0.6} L^{0.6}}{i^{0.4} S^{0.3}} \qquad (2)$$

where:

t	=	Minutes
L	=	Overland flow length, m (ft)
n	=	Manning's roughness coefficient
i	=	Rainfall rate, mm/h (in/hr)
S	=	Average slope of the overland flow area, m/m (ft/ft)
K_u	=	6.92 (.933)

Solving this equation involves iteration since both the time of concentration and rainfall intensity are unknown. When applying this equation for overland flow in turf, the n value should be quite large (e.g., 0.5). This is necessary to account for the large relative roughness resulting from water running through grass rather than over it as compared to channel flow conditions. For paved conditions an n value in the normally accepted range for smooth surfaces (e.g., 0.016) is appropriate. Hydraulic Engineering Circular Number 12 (HEC-12), entitled "Drainage of Highway Pavements" (HEC-12) illustrates the use of this equation.[6]

It is not always apparent when flow changes from overland flow to shallow concentrated flow. If a small channel or other signs of concentrated flow are not evident in the field, it is reasonable to assume a maximum overland flow length of 130 m (400 ft). As a general guideline, if the total time of concentration is less than 5 minutes, a minimum value of 5 minutes should be used for estimating the design discharge.

2.5.4. Rainfall Intensity

Rainfall intensity-frequency data are available as discussed in section 2.4. When total rainfall depths are provided the values are converted to rainfall intensity for use in the Rational formula by dividing the rainfall depth by the duration expressed in hours.

2.5.5. Drainage Area

The drainage area, in hectares, contributing flow to the point in question, can be measured on a topographic map or determined in the field. The data required to determine time of concentration and the runoff coefficient should be noted at the time of the preliminary field survey.

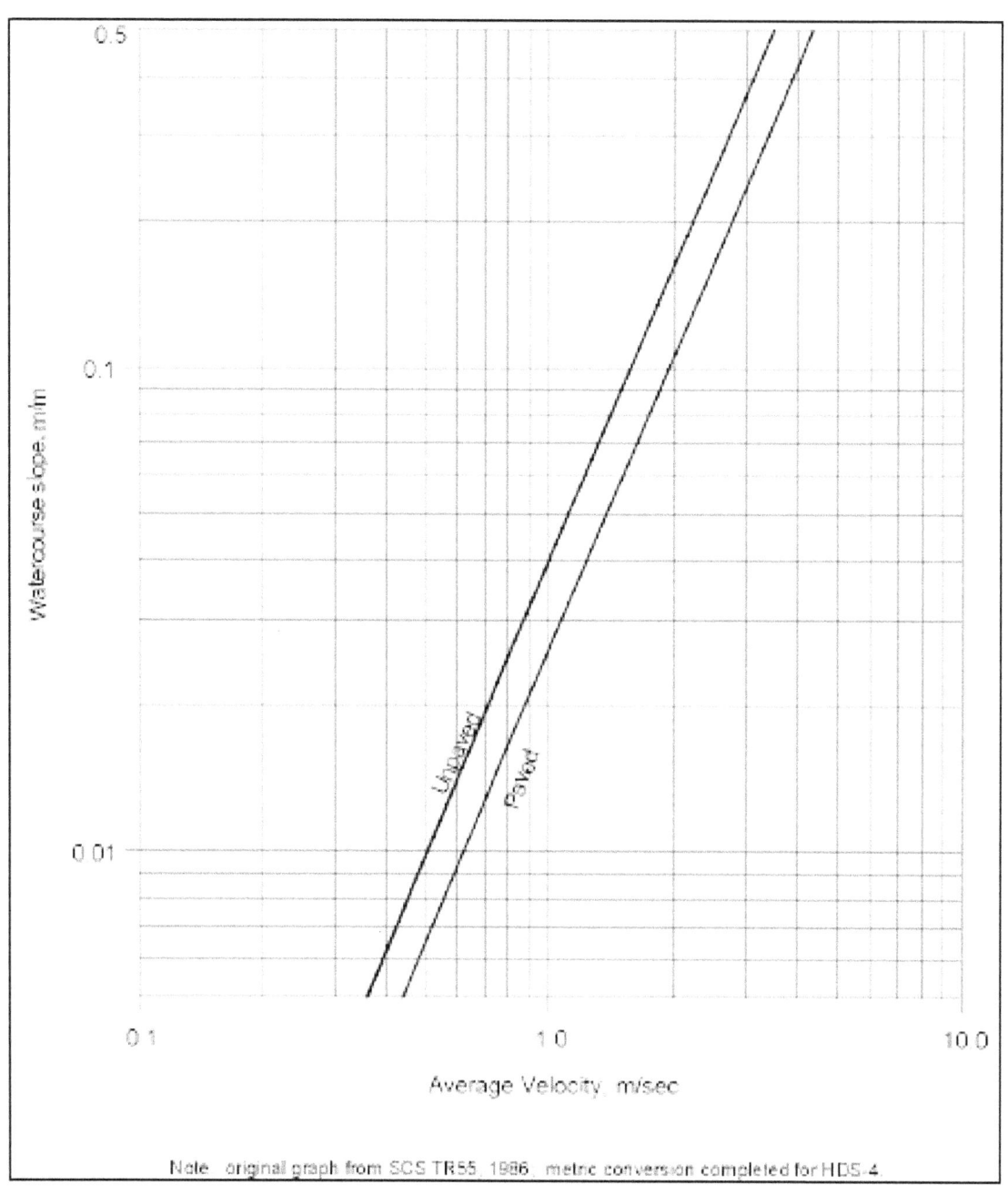

Figure 4a. Average velocities for estimating travel time for shallow concentrated flow in SI units.[7]

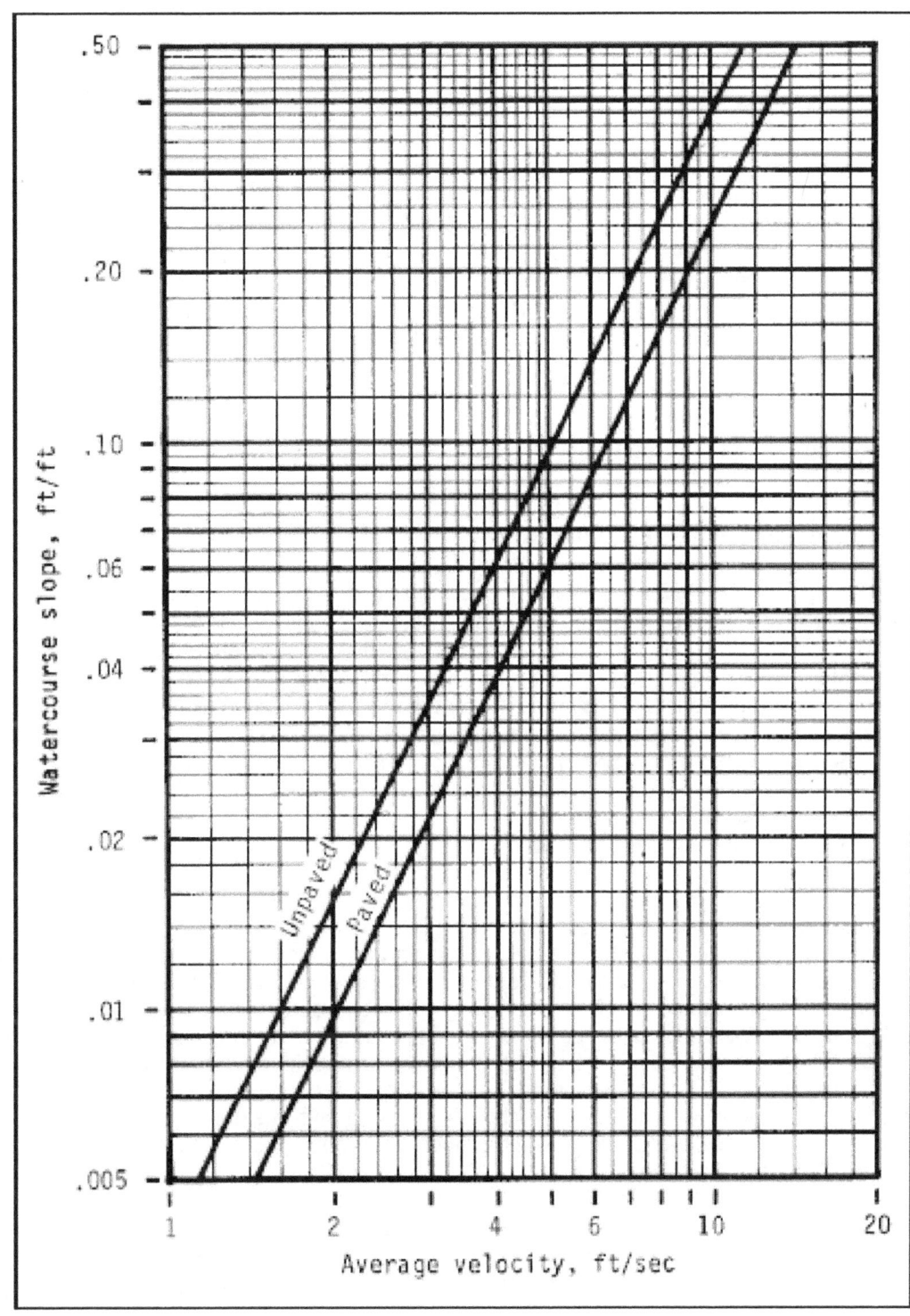

Figure 4.b. Average velocities for estimating travel time for shallow concentrated flow in English units.[7]

15

2.5.6. Computing the Design Discharge for Complex Drainage Areas

Example problem 2.2 illustrates the rational method for a simple drainage area. For other points along the channel, the design discharge is computed using the longest time of travel to the point for which the discharge is to be determined.

On some combinations of drainage areas, it is possible that the maximum rate of runoff will be reached from the higher intensity rainfall for periods less than the time of concentration for the whole area, even though only a part of the drainage area is contributing. This might occur where a part of the drainage area is highly impervious and has a short time of concentration, and another part is pervious and has a much longer time of concentration. Unless the areas or times of concentration are considerably out of balance, the accuracy of the method does not warrant checking the peak flow from only a part of the drainage area. This is particularly true for the relatively small drainage areas associated with highway pavement drainage facilities.

2.6. Regression Methods

2.6.1. Overview of Regression Methods

Regional regression equations are commonly used for estimating peak flows at ungaged sites or sites with insufficient data. Regional regression equations relate peak flow for a specified return period to the physiographic, hydrologic and meteorologic characteristics of the watershed. Regression relationships rely primarily on measured data, but may also include analytically predicted discharge estimates as part of the database used in development of a given equation. Equations have been developed for both rural and urban areas on either a state-by-state basis, or by definition of hydrophysiographic regions that may cross state boundaries.

2.6.2. Rural Regression Equations

In a series of studies by the USGS, in cooperation with the Federal Highway Administration (FHWA) and various other agencies, statewide regression equations have been developed throughout the United States. These equations permit peak flows to be estimated for return periods varying from 2 to 500 years. Typically, each state was divided into regions of similar hydrologic, meteorologic and physiographic characteristics as determined by various statistical measures. Using a combination of measured data and rainfall-runoff simulation models, long-term records of peak annual flow were synthesized for each of several watersheds in a defined region. A frequency analysis was completed on each record to define the peak discharge for a given return period. Multiple regression analysis of the peak discharge and the associated hydrologic, meteorologic and physiographic variables resulted in regression equations for peak flow determination.

The resulting set of equations, referred to as the USGS rural regression equations, were developed primarily for unregulated, natural, nonurbanized watersheds. A discussion of the accuracy of the equations and limitations in their application is provided in HDS-2.[1]

EXAMPLE PROBLEM 2.2 (SI Units)

Given: The contributing area as described in example 2.1. The weighted C is 0.35 and the channel is 125 m long on a grade of 0.5 percent.

Find: The discharge for a 10-year frequency rainfall at a storm drain inlet near the lower end of the roadside channel.

Solution: The overland flow time is obtained from equation 2. The overland flow distance is 60 m, the n value is 0.50 and S is 0.005 m/m. The rainfall intensity is initially assumed to be 55 mm/h.

$$t = 6.99 \frac{(0.5)^{0.6} (60)^{0.6}}{(55)^{0.4} (0.005)^{0.3}} = 53 \text{ min}$$

From the IDF curves (figure 3), the rainfall intensity for a 53 min duration and a 10-year return period is about 50 mm/h which is approximately equal to the initially assumed 55 mm/h for calculating t.

For the channel (125 m long) the travel time will be estimated based on average velocity and travel distance. From figure 4a, the average velocity for an unpaved surface in shallow concentrated flow is about 0.35 m/s for a slope of 0.005. For the 125 m channel the travel time is then $t_{channel}$ = (125 m)/ (0.35 m/s) = 357 s = 6 min. The total t_c = 6 + 53 = 59 min. From the rainfall charts, the 10-year rainfall intensity for 59 minutes is about 47 mm/h. The calculated discharge at the outlet of the channel is:

$$Q = \frac{0.35 \times 47 \times 0.9}{360} = 0.041 \text{ m}^3/\text{s}$$

As discussed in Section 2.5.6, this example illustrates a situation where a higher discharge may occur from the pavement area alone (with its shorter t_c) than was calculated for the entire drainage area. It would be appropriate to make an alternate calculation to evaluate this condition.

EXAMPLE PROBLEM 2.2 (English Units)

Given: The contributing area as described in example 2.1. The weighted C is 0.35 and the channel is 400 ft long on a grade of 0.5 percent.

Find: The discharge for a 10-year frequency rainfall at a storm drain inlet near the lower end of the roadside channel.

Solution: The overland flow time is obtained from equation 2. The overland flow distance is 200 ft, the n value is 0.50 and S is 0.005 ft/ft. The rainfall intensity is initially assumed to be 2.2 in/hr.

$$t = 0.933 \frac{(0.5)^{0.6}(200)^{0.6}}{(2.2)^{0.4}(0.005)^{0.3}} = 53 \text{ min}$$

From the IDF curves (figure 3), the rainfall intensity for a 53 min duration and a 10-year return period is about 2.1 in/hr which is approximately equal to the initially assumed 2.2 in/hr for calculating t.

For the channel (400 ft long) the travel time will be estimated based on average velocity and travel distance. From figure 4b, the average velocity for an unpaved surface in shallow concentrated flow is about 1.1 ft/s for a slope of 0.005. For the 400 ft channel the travel time is then $t_{channel}$ = (1.1 ft/s) = 363 s = 6 min. The total t_c = 6 + 53 = 59 min. From the rainfall charts, the 10-year rainfall intensity for 59 minutes is about 1.8 in/hr. The calculated discharge at the outlet of the channel is:

$$Q = \frac{0.35 \times 1.8 \times 2.19}{1} = 1.46 \text{ ft}^3/s$$

As discussed in Section 2.5.6, this example illustrates a situation where a higher discharge may occur from the pavement area alone (with its shorter t_c) than was calculated for the entire drainage area. It would be appropriate to make an alternate calculation to evaluate this condition.

2.6.3. Urban Regression Equations

To estimate peak discharge in urban areas, equations were developed that modify the rural peak discharge computed as described above. For a given return period, a single seven-parameter regression equation was developed for nationwide application. Equations were developed for the 2-, 5-, 10-, 25-, 50-, 100-, and 500-year events. The seven parameters used in these equations were the drainage area, the main channel slope, the 2-year rainfall intensity, the percent of the basin with reservoir or swamp storage, the basin development factor (BDF), the percentage of the basin covered by impervious area, and the equivalent rural peak discharge for the given return period.

The most significant variable found to describe the effects of urbanization was the basin development factor (BDF). The BDF, which can vary from 0 to 12, was based on a combination of several manmade changes to the drainage basin including channel improvements, channel linings, storm drains and curb and gutter streets. A complete discussion of this method is provided in USGS Water Supply Paper 2207 and HDS-2.[8,1]

2.6.4. The National Flood Frequency Program

As a result of the widespread use of various regression equations, the USGS, in cooperation with the FHWA and the Federal Emergency Management Agency (FEMA), compiled all current (as of September 1993) statewide and metropolitan area regression equations into a computer program titled the National Flood Frequency (NFF) program. A complete discussion of NFF program is provided in USGS Water Resources Investigations Report 94-4002.[9] This report summarizes the statewide regression equations for rural watersheds in each state, summarizes the applicable metropolitan area or statewide regression equations for urban watersheds, describes the NFF software for making these computations, and provides the reference information and input data needed to run the computer program. Note that typical flood hydrographs corresponding to a given peak discharge can also be estimated by procedures described in the NFF report. A summary of the NFF, and the applicability and limitations of the regression equations in the NFF, is provided in HDS-2.[1]

2.7. HYDRAIN Computer System

HYDRAIN is a comprehensive computer system for highway drainage analysis and design that was developed by FHWA and numerous state transportation agencies.[10] The HYDRAIN system includes nonproprietary computer programs to complete various hydrologic and hydraulic analysis. Many of the programs were developed to facilitate both simple and complex design analyses for problems commonly encountered by transportation agencies.

The HYDRAIN system has modules for computing rural flood frequency relationships (HYDRO) and urban relationships (HYDRA). HYDRO is composed of three different discharge volume forecasting methods: Log Pearson III, Rational Method, and multiple regression relationships. HYDRO also will develop an intensity-duration-frequency curve for a given latitude and longitude. HYDRA is a sophisticated urban storm drain model that typically would have only limited application in the small area hydrologic analyses discussed above. Nevertheless, HYDRAIN is a powerful analytical tool that may facilitate some hydrologic analyses described in this chapter, including Rational Method calculations and computation of the composite C coefficient.

(page intentionally left blank)

3. FUNDAMENTAL HYDRAULIC CONCEPTS

3.1. General

The design of drainage structures requires the use of the continuity, energy and momentum equations. From these fundamental equations other equations are derived by a combination of mathematics, laboratory experiments and field studies. These equations are used differently to analyze open-channel flow and closed conduits flowing full. A closed-conduit flowing partially full is open-channel flow. Compared to closed conduits flowing full, open-channel flow has the complexity of a free surface where the pressure is atmospheric and this free surface is controlled only by the laws of fluid mechanics. Another complexity in open-channel flow is introduced when the bed of the stream or conduit is composed of natural material such as sand, gravel, boulders or rock that is movable. In the following sections, the fundamental equations, derived equations and definitions of terms will be given. The equations and methods will not be derived. The user is referred to standard textbooks, FHWA publications and the literature cited for additional information.

Flow can be classified as: (1) uniform or nonuniform flow; (2) steady or unsteady flow; (3) laminar or turbulent flow; and (4) subcritical (tranquil) or supercritical (rapid) flow. In uniform flow, the depth, discharge, and velocity remain constant with respect to distance. In steady flow, no change occurs with respect to time at a given point. In laminar flow, the flow field can be characterized by layers of fluid, one layer not mixing with adjacent ones. Turbulent flow on the other hand is characterized by random fluid motion. Laminar flow is distinguished from turbulent flow by the use of a dimensionless number called the Reynolds Number. Subcritical flow is distinguished from supercritical flow by a dimensionless number called the Froude Number, Fr. If Fr < 1, the flow is subcritical; if Fr > 1, the flow is supercritical, and if Fr = 1, the flow is called critical. These and other terms will be more fully explained in the following sections.

3.2. Definitions

Discharge: The quantity of water moving past a given plane (cross section) in a given unit of time. Units are cubic meters per second, m^3/s (cubic feet per second, (ft^3/s)). The plane or cross section must be perpendicular to the velocity vector.

Velocity: The time rate of movement of a water particle from one point to another. The units are meters per second, m/s (feet per second (ft/s)). Velocity is a vector quantity, that is it has magnitude and direction.

Streamline: An imaginary line within the flow which is everywhere tangent to the velocity vector.

Acceleration: Acceleration is the time rate of change in magnitude or direction of the velocity vector. Units are meters per second per second, m/s^2 (feet per second per second (ft/s^2)). It is a vector quantity. Acceleration has components both tangential and normal to the streamline, the <u>tangential</u> component embodying the change in <u>magnitude</u> of the velocity, and the <u>normal</u> component reflecting a change in <u>direction</u>.

21

Local acceleration:	Local acceleration is the change in velocity (either or both magnitude and direction) with time at a given point or cross section.
Convective acceleration:	The change in velocity (either or both magnitude or direction) with distance.
Uniform flow:	In uniform flow the velocity of the flow does not change with distance. The convective acceleration is zero. Examples are flow in a straight pipe of uniform cross section flowing full or flow in a straight open channel with constant slope and all cross sections of identical form, roughness and area, resulting in a constant mean velocity.

Uniform flow conditions are rarely attained in open channels, but the error in assuming uniform flow in a channel of fairly constant slope and cross section is small in comparison to the error in determining the design discharge. |
Nonuniform flow:	In nonuniform flow the velocity of flow changes in magnitude or direction or both with distance. The convective acceleration components are different from zero. Changes occurring over long distances are classified as gradually varied flow. Changes occurring over short distances are classified as rapidly varied flow. Examples are flow around a bend or flow in expansions or contractions.
Steady flow:	In steady flow, the velocity at a point or cross section does not change with time. The local acceleration is zero.
Unsteady flow:	In unsteady flow, the velocity at a point or cross section varies with time. The local acceleration is not zero. A flood hydrograph where the discharge in a stream changes with time is an example of unsteady flow. Unsteady flow is difficult to analyze unless the time changes are small.
Laminar flow:	In laminar flow, the mixing of the fluid and momentum transfer is by molecular activity.
Turbulent flow:	In turbulent flow the mixing of the fluid and momentum transfer is related to random velocity fluctuations. The flow is laminar or turbulent depending on the value of the Reynolds Number ($Re = \rho VL/\mu$), which is a dimensionless ratio of the inertial forces to the viscous forces. Here ρ and μ are the density and dynamic viscosity of the fluid, V is the fluid velocity, and L is a characteristic dimension, usually the depth (or the hydraulic radius) in open-channel flow. In laminar flow, viscous forces are dominant and Re is relatively small. In turbulent flow, Re is large; that is, inertial forces are very much greater than viscous forces. Turbulent flows are predominant in nature. Laminar flow occurs very infrequently in open-channel flow.

Open-channel flow: Open-channel flow is flow with a free surface. Closed-conduit flow or flow in culverts is open-channel flow if they are not flowing full and there is a free surface.

Froude Number: The Froude Number $Fr = \dfrac{V}{\sqrt{gL}}$ is the ratio of inertial forces to gravitational forces. The Froude Number is also the ratio of the flow velocity V to the celerity $C = \sqrt{gL}$ of a small gravity wave in the flow.

Subcritical flow: Open-channel flow's response to changes in channel geometry depends upon the depth and velocity of the flow. Subcritical flow (or tranquil flow) occurs on mild slopes where the flow is deep with a low velocity and has a Froude Number less than 1. In subcritical flow, the boundary condition (control section) is always at the downstream end of the flow reach.

Supercritical flow: Supercritical flow occurs on steep slopes where the flow is shallow with a high velocity and has a Froude Number greater than 1. In supercritical flow, the boundary condition (control section) is always at the upstream end of the flow reach.

Critical flow: When the Froude Number equals 1, the flow is critical and surface disturbances remain stationary in the flow.

Pressure flow: Flow in a closed conduit or culvert that is flowing full with water in contact with the total enclosed boundary and under pressure.

Closed-conduit flow: Flow in a pipe, culvert, etc. where there is a solid boundary on all four sides. Examples are pipes, culverts, and box culverts. Flow conditions in a closed conduit may occur as gravity full flow, full-pressure flow, or partly full (open-channel flow).

Alluvial channel: Flow in an open channel where the bed is composed of material that has been deposited by the flow.

Hydraulic radius: The hydraulic radius is a length term used in many of the hydraulic equations that is determined by dividing the flow area by the length of the cross section in contact with the water (wetted perimeter). The hydraulic radius is in many of the equations to help take into account the effects of the shape of the cross section on the flow. For example, the hydraulic radius for a circular pipe flowing full is equal to the diameter of the pipe divided by four (D/4).

One dimensional flow: A method of analysis where changes in the flow variables (velocity, depth, etc.) occur primarily in the longitudinal direction. Changes of flow variables in the other two dimensions are small and are neglected.

Two-dimensional flow: A method of analysis where the accelerations can occur in two directions (along and across the flow).

Three-dimensional flow: The flow variables can change in all three dimensions, along, across, and in the vertical.

3.3. Basic Principles

3.3.1. Introduction

The basic equations of flow are continuity, energy and momentum. They are derived from the laws of (1) the conservation of mass; (2) the conservation of energy; and (3) the conservation of linear momentum, respectively. Conservation of mass is another way of stating that (except for mass-energy interchange) matter can neither be created nor destroyed. The principle of conservation of energy is based on the first law of thermodynamics which states that energy must at all times be conserved. The principle of conservation of linear momentum is based on Newton's second law of motion which states that a mass (of fluid) accelerates in the direction of and in proportion to the applied forces on the mass.

Analysis of flow problems are much simplified if there is no acceleration of the flow or if the acceleration is primarily in one direction (one-dimensional flow), the accelerations in other directions being negligible. However, a very inaccurate analysis may occur if one assumes accelerations are small or zero when in fact they are not. The concepts given in this manual assume one-dimensional flow. Only the equations will be given. The user is referred to standard fluid mechanics texts or "Highways in the River Environment" (HIRE) for their derivations.[11]

3.3.2. Continuity Equation

The continuity equation is based on conservation of mass. For steady flow of incompressible fluids it is:

$$V_1 A_1 = V_2 A_2 = Q = VA \tag{3}$$

where:

V	$=$	Average velocity in the cross section perpendicular to the area, m/s (ft/s)
A	$=$	Area perpendicular to the velocity, m^2 (ft^2)
Q	$=$	Volume flow rate or discharge, m^3/s (ft^3/s)

Equation 3 is applicable when the fluid density is constant, the flow is steady, there is no significant lateral inflow or seepage (or they are accounted for) and the velocity is perpendicular to the area (figure 5).

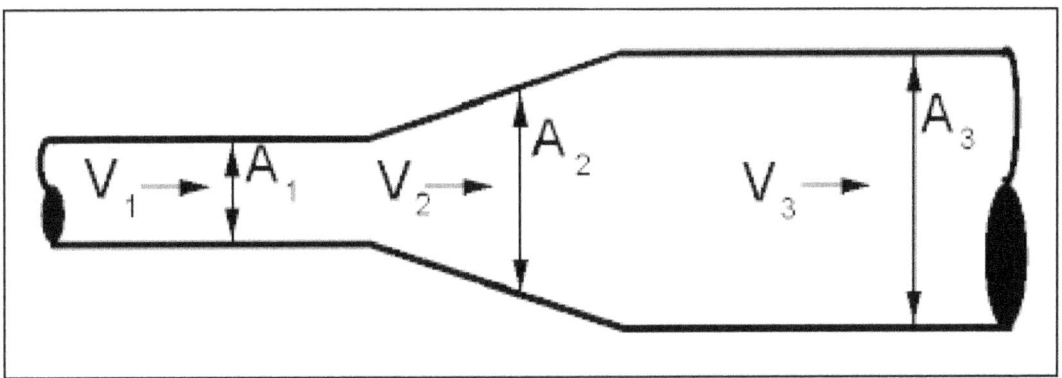

Figure 5. Sketch of continuity concept.

3.3.3. Energy Equation

The energy equation is derived from the first law of thermodynamics which states that energy must be conserved at all times. The energy equation is a scalar equation. For steady incompressible flow it is:

$$\alpha_1 \frac{V_1^2}{2g} + \frac{p_1}{\gamma} + Z_1 = \alpha_2 \frac{V_2^2}{2g} + \frac{p_2}{\gamma} + Z_2 + h_L \qquad (4)$$

where:

α	=	Kinetic energy correction factor
V	=	Average velocity in the cross section, m/s (ft/s)
g	=	Acceleration of gravity, 9.81 m/s^2 (32.2 ft/s^2)
p	=	Pressure, N/m^2 or Pa (lbs/ft^2)
γ	=	Unit weight of water, 9,800 N/m^3 (62.4 lbs/ft^3) at 15°C (59°F)
Z	=	Elevation above a horizontal datum, m (ft)
h_L	=	Headloss due to friction and form losses, m (ft)
A	=	Area of the cross section, m^2 (ft^2)

The kinetic energy correction factor α is to correct for the velocity distribution across the flow. This allows the use of the average velocity (V) rather than the point velocity (v). It is given by the following equation:

$$\alpha = \frac{1}{V^3 A} \int_A v^3 \, dA \qquad (5)$$

25

EXAMPLE PROBLEM 3.1 (SI Units)

Given: A storm drain flowing full transitions from 0.7 m to 1.0 m diameter pipe. Determine the average velocity in each section of pipe for a discharge 0.5 m³/s.

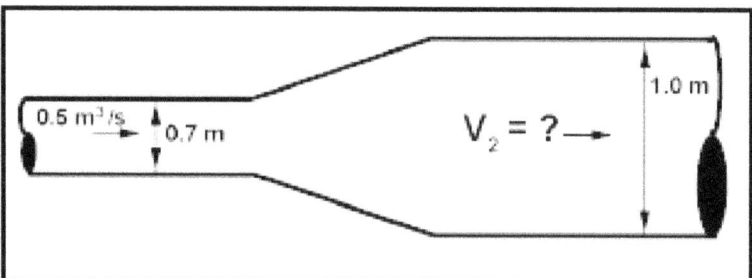

Find:

(a)   Velocity at section 1 (0.7 m pipe)
(b)   Velocity at section 2 (1.0 m pipe)

Solution:

Since the discharge at the beginning of the pipe must equal the discharge at the end of the pipe, the continuity equation can be used:

Basic equation: $Q = VA$ Rearrange to get $V = \dfrac{Q}{A}$ For a circular pipe: $A = \dfrac{\pi D^2}{4}$

At cross section 1:

$$V = \left[\frac{0.5 \, m^3 / s}{\dfrac{\pi (0.7m)^2}{4}} \right] = 1.30 \, m/s$$

At cross section 2:

$$V = \left[\frac{0.5 \, m^3 / s}{\dfrac{\pi (1m)^2}{4}} \right] = 0.64 \, m/s$$

EXAMPLE PROBLEM 3.1 (English Units)

Given: A storm drain flowing full transitions from 24- to 36-inch diameter pipe. Determine the average velocity in each section of pipe for a discharge 10 ft³/s.

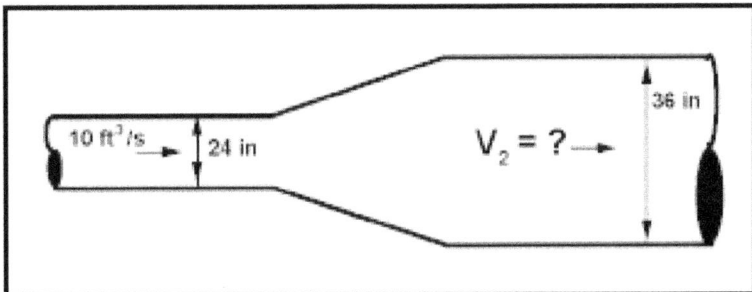

Find:

(a)   Velocity at section 1 (24-inch pipe)
(b)   Velocity at section 2 (36-inch pipe)

Solution:

Since the discharge at the beginning of the pipe must equal the discharge at the end of the pipe, the continuity equation can be used:

Basic equation: $Q = VA$ Rearrange to get $V = \dfrac{Q}{A}$ For a circular pipe: $A = \dfrac{\pi D^2}{4}$

At cross section 1:

$$V = \left[\frac{10\,\text{ft}^3/\text{s}}{\dfrac{\pi(2\,\text{ft})^2}{4}}\right] = 3.18\,\text{ft}/\text{s}$$

At cross section 2:

$$V = \left[\frac{10\,\text{ft}^3/\text{s}}{\dfrac{\pi(3\,\text{ft})^2}{4}}\right] = 1.42\,\text{ft}/\text{s}$$

where:

v $=$ Velocity at a point or average in a vertical, m/s (ft/s)

Note that even with a very nonuniform velocity distribution across a section the correction is only 10 percent. Consequently, the kinetic energy correction factor is normally equal to 1.0.

The energy grade line (EGL) represents the total energy at any given cross section, defined as the sum of the three components of energy represented on each side of equation 4. These components of energy are often referred to as the velocity head, pressure head, and elevation head. The hydraulic grade line (HGL) is below the EGL by the amount of the velocity head, or is the sum of just the pressure head and the elevation head. The application of the energy equation in open-channel and pressure flow is illustrated in figures 6 and 7.

3.3.4. Momentum Equation

The momentum equation is derived from Newton's second law which states that the summation of all external forces on a system is equal to the change in momentum. In the x-direction for steady flow with constant density it is:

$$F_x = \rho Q (\beta_2 V_{x2} - \beta_1 V_{x1}) \tag{6}$$

where:

F_x	$=$	Forces in the x direction, N (lbs)
ρ	$=$	Density, 1000 Kg/m^3 (1.94 slugs/ft^3)
β	$=$	Momentum coefficient
Q	$=$	Volume flow rate or discharge, m^3/s (ft^3/s)
V	$=$	Velocity in the x direction, m/s (ft/s)

The momentum coefficient corrects for the velocity distribution across the flow. Again, this allows the use of the average velocity (V) rather than the point velocity (v). It is given by

$$\beta = \frac{1}{V^2 A} \int_A v^2 \, dA \tag{7}$$

The momentum coefficient is normally assumed to be 1.0 since a very nonuniform velocity distribution across a section would only require a correction less than 10 percent. The momentum equation is a vector equation and similar equations are used for the y and z directions.

3.3.5. Hydrostatics

When the only forces acting on the fluid are pressure and fluid weight, the differential equation of motion in an arbitrary direction x is

$$\frac{\partial}{\partial x}\left(\frac{p}{\gamma} + z\right) = \frac{a_x}{g} \tag{8}$$

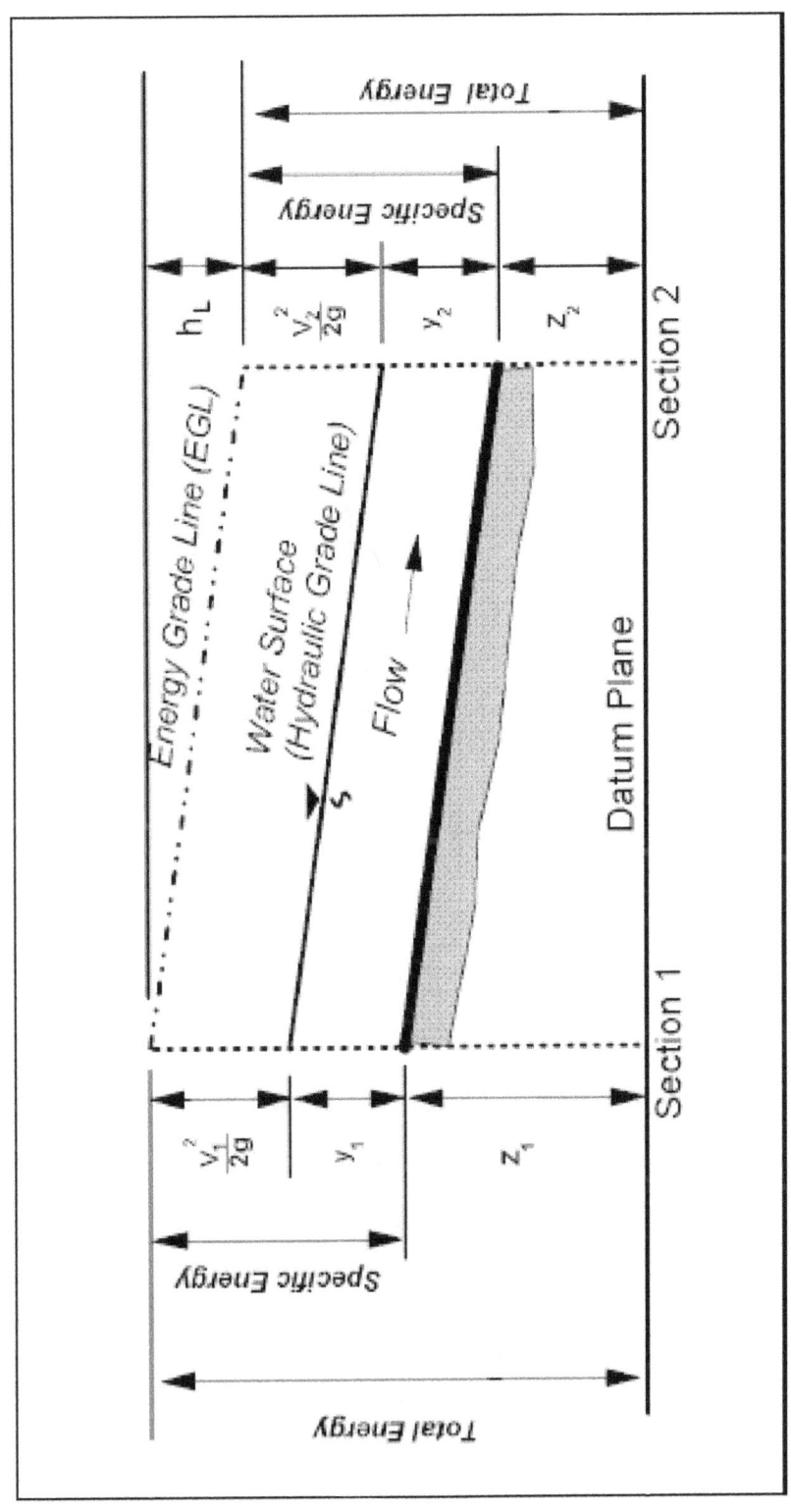

Figure 6. Sketch of energy concept for open-channel flow.

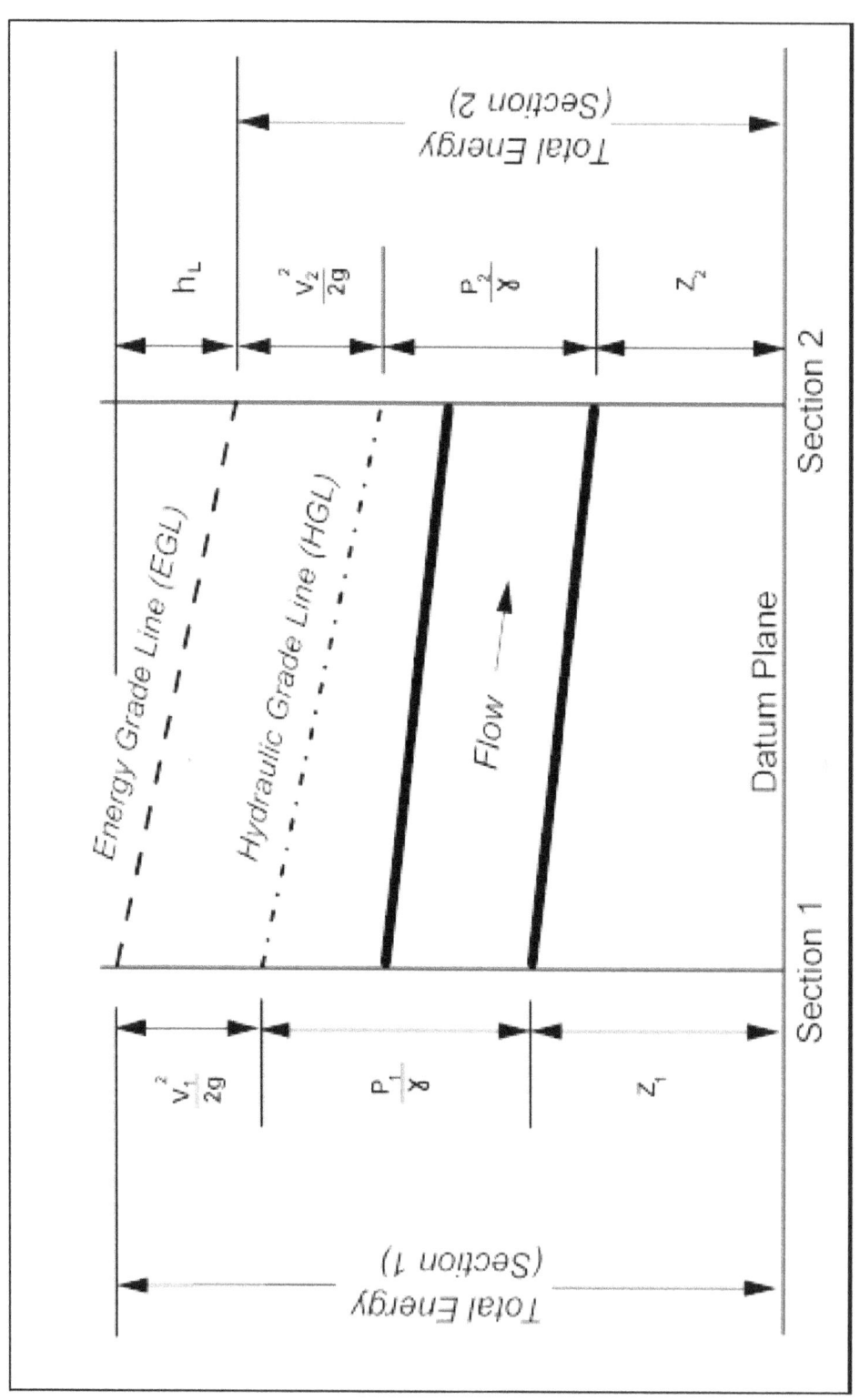

Figure 7. Sketch of energy concept for pressure flow.

EXAMPLE PROBLEM 3.2 (SI Units)

Given: The velocity of the upstream end of a rectangular channel 1 m wide is 3.0 m/s, and the flow depth is 2.0 m. The depth at the downstream end is 1.7 m. The elevation at section 1 is 500 m and at 2 is 499.90 m. Determine the headloss due to friction. Assume the kinetic energy correction factor is 1.0.

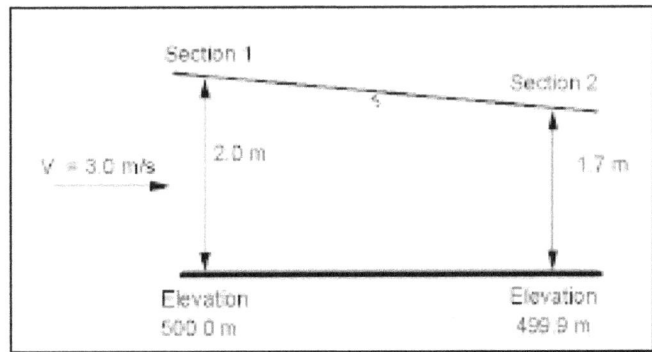

Find:
(a)   Headloss (h_L)

Solution:
Step 1: Use the continuity equation to find the velocity at section 2.

$Q = VA$

$$V_2 = \frac{Q}{A_2} = \frac{(3.0\,m/s)(2.0\,m \times 1.0\,m)}{1.7\,m \times 1.0\,m} = 3.53\,m/s$$

Step 2: Use the energy equation to find the headloss, h_L

$$\frac{V_1^2}{2g} + Y_1 + Z_1 = \frac{V_2^2}{2g} + Y_2 + Z_2 + h_L$$

$$\frac{(3.0)^2}{2(9.81)} + 2.0 + 500 = \frac{(3.53)^2}{2(9.81)} + 1.7 + 499.9 + h_L$$

$$h_L = (0.46 + 2 + 500) - (0.63 + 1.7 + 499.9) = 0.23\,m$$

EXAMPLE PROBLEM 3.2 (English Units)

Given: The velocity of the upstream end of a rectangular channel 3 ft wide is 10.0 ft/s, and the flow depth is 6.5 ft. The depth at the downstream end is 6 ft. The elevation at section 1 is 1640.5 ft and at 2 is 1640.0 ft. Determine the headloss due to friction. Assume the kinetic energy correction factor is 1.0.

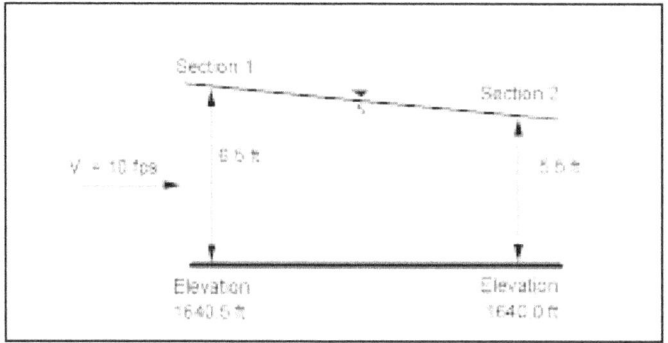

Find:

(a)   Headloss (h_l)

Solution:

Step 1: Use the continuity equation to find the velocity at section 2.

$Q = VA$

$$V_2 = \frac{Q}{A_2} = \frac{(10\,ft/s)(6.5\,ft)(3\,ft)}{(5.5\,ft)(3.0\,ft)} = 11.82\,ft/s$$

Step 2: Use the energy equation to find the headloss, h_L

$$\frac{V_1^2}{2g} + Y_1 + Z_1 = \frac{V_2^2}{2g} + Y_2 + Z_2 + h_L$$

$$\frac{(10)^2}{2(32.2)} + 6.3 + 1640.3 = \frac{(11.82)^2}{2(32.2)} + 5.3 + 1640.0 + h_L$$

$$h_L = (1.55 + 6.5 + 1640.3) - (2.17 + 5.5 + 1640.0) = 0.88\,ft$$

EXAMPLE PROBLEM 3.3 (SI Units)

Given: For a bridge widening project, an existing city watermain must be relocated. The water main is 300 mm in diameter and carries 0.142 m³/s. Relocation of the water main will require a 45° bend in the pipe. The pressure in the pipe at the location of the bend is 689 464 N/m². Determine the forces that an anchor on the pipe at the bend needs to withstand.

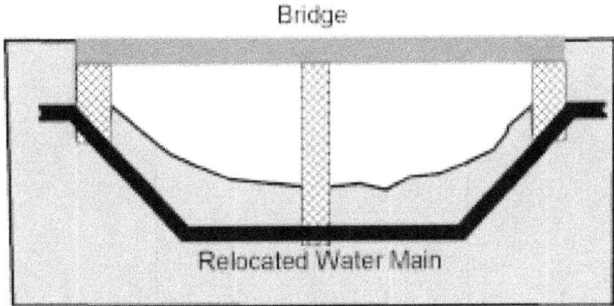

Bridge

Relocated Water Main

Find:

(a)　Forces on bend

Solution:

(a)　Velocity in pipe

$$A = \frac{\pi D^2}{4} = \frac{\pi (.3)^2}{4} = 0.071 m^2$$

$$Q = VA \rightarrow V = \frac{Q}{A} = \frac{0.142\, m^3/s}{0.071 m^2} = 2.0\, m/s$$

33

(b)   Use the momentum equation to find the forces on the bend in the x- and y-directions. First, draw a diagram of the bend and label the forces. (Note: sign convention is important in drawing a force diagram).

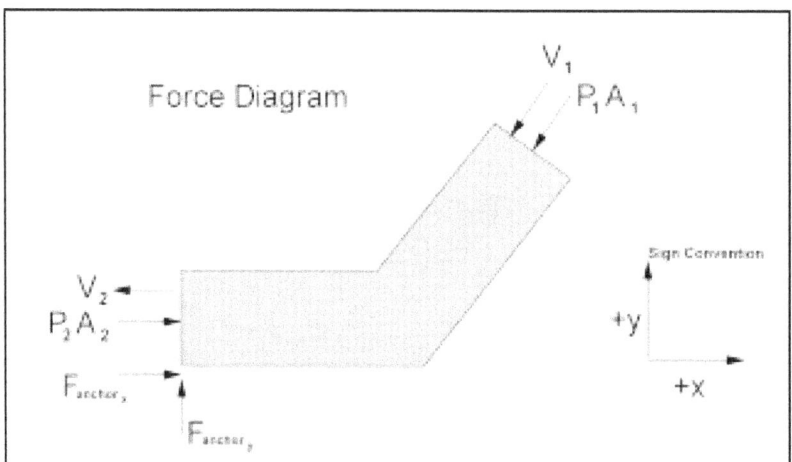

The momentum equation states

$$F_x = \rho Q (V_{2x} - V_{1x})$$

The forces acting on the pipe include pressure on either side of the pipe and the resisting of the anchor.

$$F_{anchor_x} + P_{2x} A_{2x} - P_{1x} A_{1x} \cos 45 = \rho Q (V_{2x} - V_1 \cos 45)$$

First, determine the forces in the x direction (assuming no change in pressure through the bend):

$$F_{anchor_x} + (689\,464)(0.071) - (689\,464)(.071) \cos 45 = 1000(0.142)(-2.0 + 2.0 \cos 45)$$

$$F_{anchor_x} + 48\,952 - 34\,614 = -83$$

$$F_{anchor_x} = -14\,421\,Nm$$

34

Next, determine the forces in the y direction:

$$F_y = \rho Q (V_{2y} - V_{1y})$$

$$F_{anchor_y} + P_{2y} A_{2y} - P_{1y} A_{1y} = \rho Q (V_{2y} - V_{1y})$$

$$F_{anchor_y} + 0 - 689\,464(0.071) \, Sin\,45 = 1000(0.142)(0 + 2.0\,Sin\,45)$$

$$F_{anchor_y} - 34\,614 = 201$$

$$F_{anchor_y} = 34\,815\,N$$

$$F_{total} = \sqrt{(14\,421)^2 + (34\,814)^2}$$

$F_{total} = 37\,683\,N$ acting at about 113 degrees clockwise from the + X axis

This is equivalent to the weight of two or three automobiles sitting atop the anchor!

EXAMPLE PROBLEM 3.3 (English Units)

Given: For a bridge widening project, an existing city watermain must be relocated. The water main is 12 inches in diameter and carries 5 ft³/s. Relocation of the water main will require a 45° bend in the pipe. The pressure in the pipe at the location of the bend is 100 lb/in². Determine the forces that an anchor on the pipe at the bend needs to withstand.

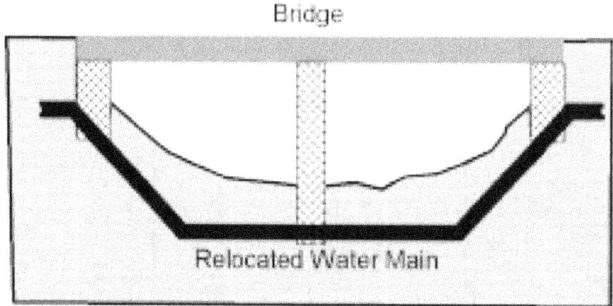

Bridge

Relocated Water Main

Find:

(a)   Forces on bend

Solution:

(a)   Velocity in pipe

$$A = \frac{\pi D^2}{4} = \frac{\pi (1)^2}{4} = 0.79 \, ft^2$$

$$Q = VA \rightarrow V = \frac{Q}{A} = \frac{5 \, ft^3 / s}{0.79 \, ft^2} = 6.33 \, ft / s$$

(b)   Use the momentum equation to find the forces on the bend in the x- and y-directions. First, draw a diagram of the bend and label the forces. (Note: sign convention is important in drawing a force diagram).

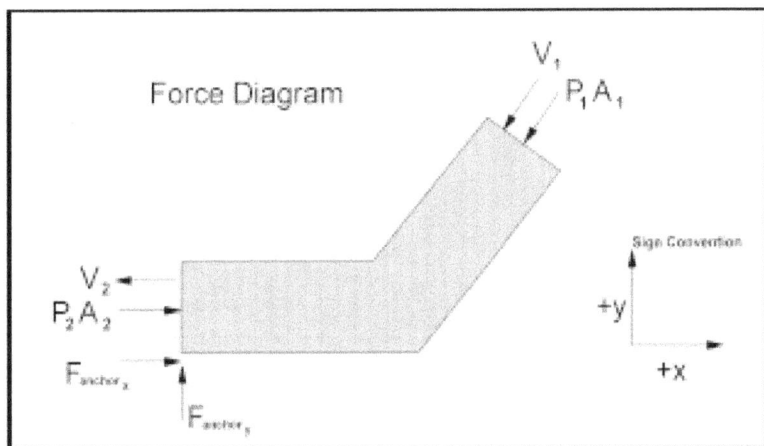

The momentum equation states

$$F_x = \rho Q (V_{2x} - V_{1x})$$

The forces acting on the pipe include pressure on either side of the pipe and the resisting of the anchor.

$$F_{anchor_x} + P_{2x} A_{2x} - P_{1x} A_{1x} \cos 45 = \rho Q (V_{2x} - V_1 \cos 45)$$

First, determine the forces in the x direction (assuming no change in pressure through the bend):

$$F_{anchor_x} + 100 lb/in^2 \left(\frac{144 in^2}{ft^2} \right) (0.79 ft^2) - 100 lb/in^2 \left(\frac{144 in^2}{ft^2} \right) (0.79 ft^2) \cos 45°$$

$$= \left(1.94 \frac{lbs^2}{ft^4} \right) \left(\frac{5 ft^3}{S} \right) (-6.33 ft/s + 6.33 ft/s \cos 45°)$$

$$F_{anchor_x} + 11,376 lb - 8,044 lb = -18 lb$$

$$F_{anchor_x} = -3,332 lb - 18 lb = -3,350 lb$$

Next, determine the forces in the y direction:

$$F_y = \rho Q (V_{2y} - V_{1y})$$

$$F_{anchor_y} + P_{2y} A_{2y} - P_{1y} A_{1y} = \rho Q (V_{2y} - V_{1y})$$

$$F_{anchor_y} + 0 - 100\text{lb}/\text{in}^2 \left(\frac{144\ \text{in}^2}{\text{ft}^2} \right) (0.79\ \text{ft}^2) \text{Sin } 45^\circ = \left(1.94 \frac{\text{lbs}^2}{\text{ft}^4} \right) (5\ \text{ft}^3/\text{s})(0 + 6.33\ \text{ft}/\text{s})(\text{Sin } 45)$$

$$F_{anchor_y} - 8,044\ \text{lb} = 43\ \text{lb}$$

$$F_{anchor_y} = 8,087\ \text{lb}$$

$$F_{total} = \sqrt{(3,350)^2 + (8,087)^2}$$

F_{total} = 8,753 lb acting at about 113 degrees clockwise from the + X axis

This is equivalent to the weight of two or three automobiles sitting atop the anchor!

In steady uniform flow (and for zero flow), the acceleration is zero and we obtain the equation of hydrostatics

$$\frac{p}{\gamma} + Z = \text{Constant} \tag{9}$$

However, when there is acceleration, the piezometric head term $(p/\gamma+Z)$ varies in the flow field. That is, the piezometric head is not constant in the flow. This is illustrated in figure 8. In figure 8a the pressure at the bed is hydrostatic and equal to γy_o whereas in curvilinear flow (figure 8b) the pressure is larger than γy_o because of the acceleration resulting from a change in direction.

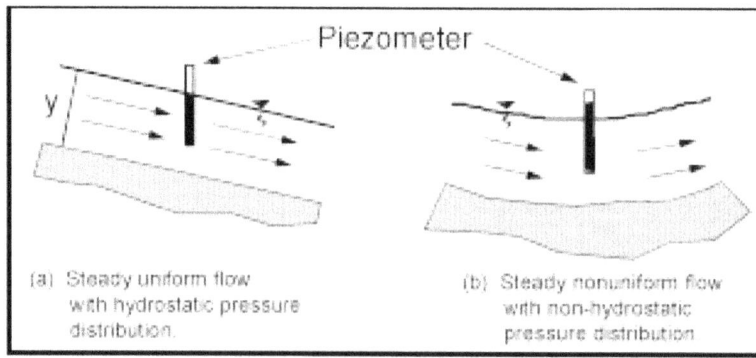

(a) Steady uniform flow with hydrostatic pressure distribution.

(b) Steady nonuniform flow with non-hydrostatic pressure distribution

Figure 8. Pressure distribution in steady uniform and in steady nonuniform flow.

In general, when fluid acceleration is small (as in gradually varied flow) the pressure distribution is considered hydrostatic. However, for rapidly varying flow where the streamlines are converging, expanding or have substantial curvature (curvilinear flow), fluid accelerations are not small and the pressure distribution is not hydrostatic.

In equation 9 the constant is equal to zero for gage pressure at the free surface of a liquid, and for flow with hydrostatic pressure throughout (steady, uniform flow or gradually varied flow) it follows that the pressure head p/γ is equal to the vertical distance below the free surface. In sloping channels with steady uniform flow, the pressure head p/γ at a depth y below the surface is equal to

$$\frac{p}{\gamma} = y \cos \theta \qquad (10)$$

Note that y is the depth (perpendicular to the water surface) to the point, as shown in figure 9. For most channels, θ is small and $\cos \theta \approx 1$ for channels with slopes less than ten percent.

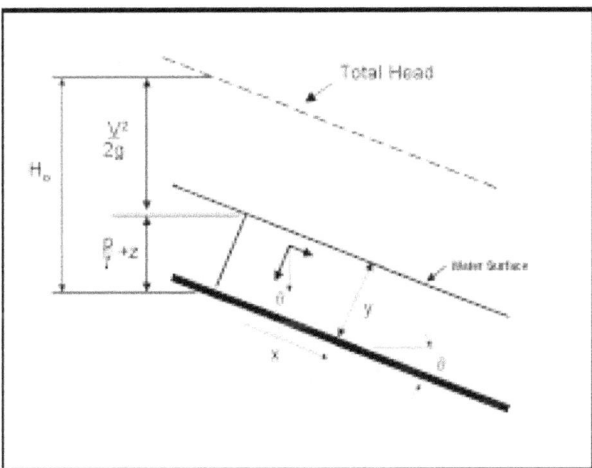

Figure 9. Pressure distribution in steady uniform flow on steep slopes.

3.4. Weirs and Orifices

3.4.1. Weirs

A weir is typically a notch of regular shape (rectangular, square, or triangular), with a free surface. The edge or surface over which the water flows is called the crest. A weir with a crest where the water springs free of the crest at the upstream side is called a sharp crested weir. If the water flowing over the weir does not spring free and the crest length is short, the weir is called a not sharp crested weir, round edge weir, or suppressed weir. If the weir has a horizontal or sloping crest sufficiently long in the direction of flow that the flow pressure distribution is hydrostatic it is called a broad crested weir (figure 10). As with orifices, weirs can be used to measure water flow. Strictly speaking a sharp crested weir, for measurement purposes, must be aerated on the downstream side and the pressure on the nape downstream be atmospheric. Examples of weir flow that are of interest to the highway engineer are, flow-over approach embankment and flow spilling through curb inlets.

39

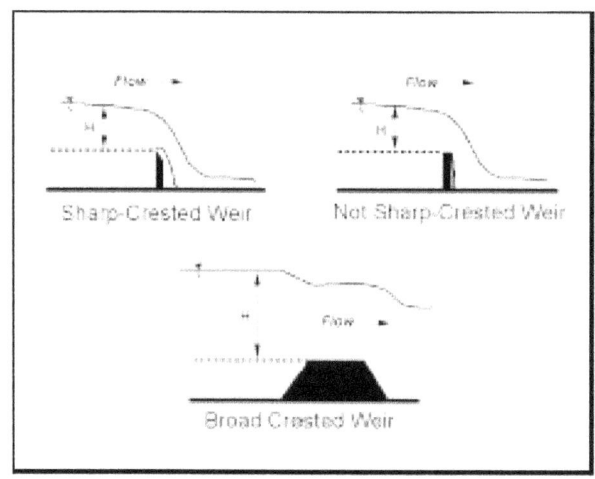

Figure 10. Weir types.

The discharge across a weir (sharp-crested or broad-crested) is:

$$Q = C_b \, L \, H^{3/2} \tag{11}$$

where:

Q   =   Discharge, m^3/s (ft^3/s)
C_D =   Coefficient of discharge for weirs, sharp edge or broad crested
L   =   Flow length across the weir, m (ft)
H   =   The head on the weir, m (ft). The depth of flow above the weir crest upstream of the weir (typically measured a distance of about 2.5H upstream of the weir).

Coefficients of discharge are given in most handbooks (e.g., references 12 and 13)) for the different types of weirs or flow conditions. Note that since C_D has units of $\sqrt{g}$, C_D values tabulated in English units must be converted to metric units by multiplying by the factor $\sqrt{9.81} / \sqrt{32.2}$ or 0.552. Correction factors are also available if the weir is submerged (tailwater above the weir, see reference 14). As long as the tailwater is less than critical depth, submergence is not a factor.

3.4.2. Orifices

An orifice is an opening with a regular shape (circular, square or rectangular) through which water flows in contact with the total perimeter. If the opening is flowing only partially full, the orifice becomes a weir. An orifice with a sharp upstream edge is called a sharp-edged orifice. If the jet of water from the orifice discharges into the air, it is called a free discharge. If it discharges under water, it is called a submerged orifice. Orifices are common fluid discharge measuring devices (figure 11) but orifice type flow occurs under other circumstances where headloss, backwater etc. needs to be determined. Examples of orifice flows of interest to highway engineers are flow through bridges when they are overtopped, flow-through culvert inlets, curb inlets flowing full, etc. When a bridge is overtopped the flow through the bridge is orifice flow, but the flow over the bridge is weir flow.

EXAMPLE PROBLEM 3.4 (SI Units)

Given: During a flood, water overtops a roadway embankment at a sag in the roadway profile. Determine the amount of flow over the road and its velocity if the inundated roadway length = 130 m and C_D = 3.1 (English units). The flow area was calculated to be 390 m^2 based on a high-water mark on a tree and the roadway profile.

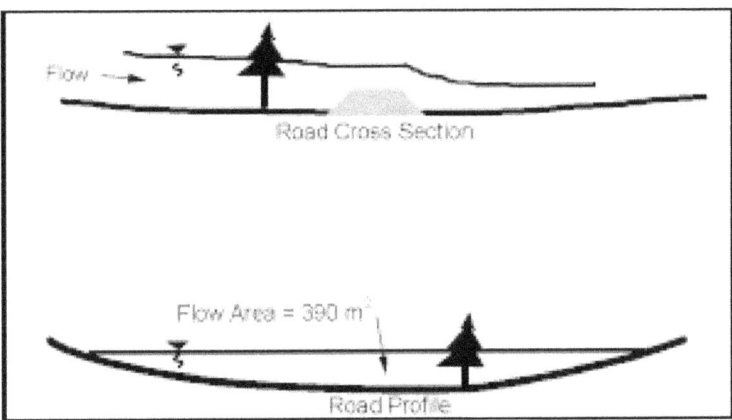

Find:

(1)   Discharge across the road
(2)   Velocity across the road

Solution:

Use the broad-crested weir equation and the continuity equation since the road acts as a weir.

$Q = C_b L H^{3/2}$ and $Q = VA$

Since the flow depth changes across the length of the road, use the hydraulic depth for H (area/topwidth)

$$Q = (3.1 \times 0.552)\,(130) \left(\frac{390}{130} \right)^{3/2} = 1156 \text{ m}^3/\text{s}$$

Find the velocity of the flow from the continuity equation.

$Q = VA$

$$V = \frac{1156 \text{ m}^3/\text{s}}{390 \text{ m}^2} = 3.0 \text{ m/s}$$

EXAMPLE PROBLEM 3.4 (English Units)

Given: During a flood, water overtops a roadway embankment at a sag in the roadway profile. Determine the amount of flow over the road and its velocity if the inundated roadway length = 425 ft and C_D = 3.1 (English units). The flow area was calculated to be 4,250 ft^2 based on a high-water mark on a tree and the roadway profile.

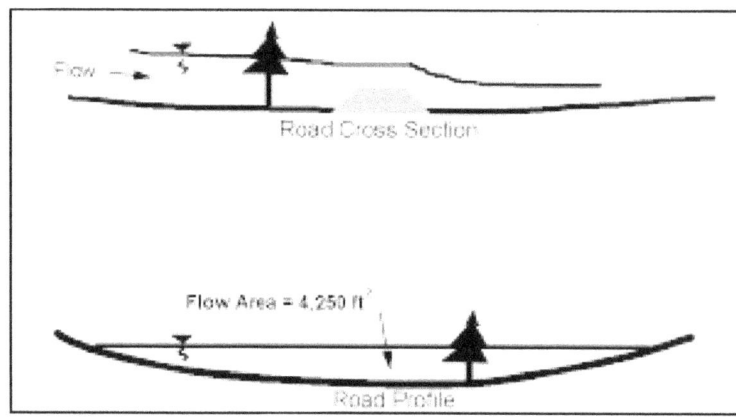

Find:

(1)   Discharge across the road
(2)   Velocity across the road

Solution:

Use the broad-crested weir equation and the continuity equation since the road acts as a weir.

$Q = C_D L H^{3/2}$ and $Q = VA$

Since the flow depth changes across the length of the road, use the hydraulic depth for H (area/topwidth)

$$Q = (3.1)(425)\left(\frac{4,250}{425}\right)^{3/2} = 41,663 \, ft^3/s$$

Find the velocity of the flow from the continuity equation.

$Q = VA$

$$V = \frac{41,663 \, ft^3/s}{4,250 \, ft^2} = 9.8 \, ft/s$$

42

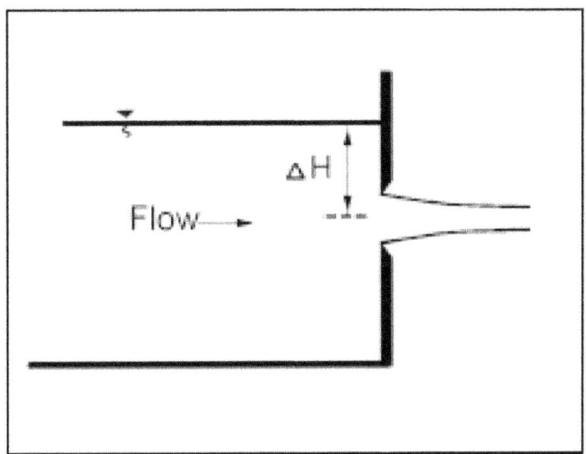

Figure 11. Orifice.

The discharge through an orifice is:

$$Q = C_D A \sqrt{2g\Delta H} \qquad (12)$$

where:

Q	=	Discharge, m^3/s (ft^3/s)
C_D	=	Coefficient of discharge
A	=	Area of the orifice, m^2 (ft^2)
g	=	Acceleration of gravity = 9.81 m/s^2 (32.2 ft/s^2)
ΔH	=	The difference in head across the orifice, m (ft)

Coefficients of discharge are given in most handbooks.[12,13] For an unsubmerged orifice, the difference in head across the orifice is measured from the centerline of the orifice to the upstream water surface. For a submerged orifice, the difference in head is measured from the upstream water surface to the downstream water surface.

(page intentionally left blank)

4. OPEN-CHANNEL FLOW

4.1. Introduction

Open-channel flow is more complex than closed-conduit flow flowing full because the water surface is determined by the mechanics of motion. In addition, if the bottom boundary is movable (alluvial boundary) another complexity is introduced. When the channel is mobile, the resistance to flow is a function of the flow.

In this chapter the concepts and equations for the simplest flow condition (steady, uniform flow) will be described, as well as the bedform conditions that occur in an alluvial channel. Flow conditions and equations for solving problems of increasing flow complexity will be given. The one-dimensional method will be used in the descriptions of the equations.

4.2. Alluvial Channel Flow

4.2.1. Alluvial Channels

Alluvial channels are channels formed in material that has been and can be transported by the flow. They are commonly made up of bed material composed of sand-, gravel-, and cobble-sized material. These materials are important in drainage design because they are the most common materials encountered and they affect resistance to flow and erosion. Concrete channels and culverts may have an alluvial boundary because of deposition of bed material in the invert.

4.2.2. Bedforms in Sand Channels

The predominant material in sand-bed streams ranges from coarse silt to sand. There may be finer or coarser material in the bed, but the dominant size will be sand (50 percent or more). In sand-bed streams, the bed material is easily eroded and continually being moved and shaped by the flow. The interaction between the flow of the water-sediment mixture and the sand bed creates different bed configurations which change the resistance to flow, velocity, water surface elevation, and sediment transport. Consequently, it is necessary to understand what bedforms will be present so that the resistance to flow can be estimated and flood stages, depth of flow, and water surface profiles can be computed in order to design drainage channels.

4.2.3. Flow Regime

Flow in alluvial channels is divided into two regimes separated by a transition zone. Forms of bed roughness in sand channels are shown in figure 12. The flow regimes are:

- The lower flow regime, where resistance to flow is large and sediment transport is small. The bedform is either ripples or dunes or some combination of the two. Water-surface undulations are out of phase with the bed surface, and there is a relatively large separation zone downstream from the crest of each ripple or dune. The velocity of the downstream movement of the ripples or dunes depends on their height and the velocity of the grains moving up their backs.

45

- The transition zone, where the bed configuration may range from that typical of the lower flow regime to that typical of the upper flow regime, depending mainly on antecedent conditions. If the antecedent bed configuration is dunes, the depth or slope can be increased to values more consistent with those of the upper flow regime without changing the bedform; or, conversely, if the antecedent bed is plane, depth and slope can be decreased to values more consistent with those of the lower flow regime without changing the bedform.

  Resistance to flow and sediment transport also have the same variability as the bed configuration in the transition. This phenomenon can be explained by the changes in resistance to flow and, consequently, the changes in depth and slope as the bedform changes.

- The upper flow regime, in which resistance to flow is small and sediment transport is large. The usual bedforms are plane bed or antidunes. The water surface is in phase with the bed surface except when an antidune breaks, and normally the fluid does not separate from the boundary.

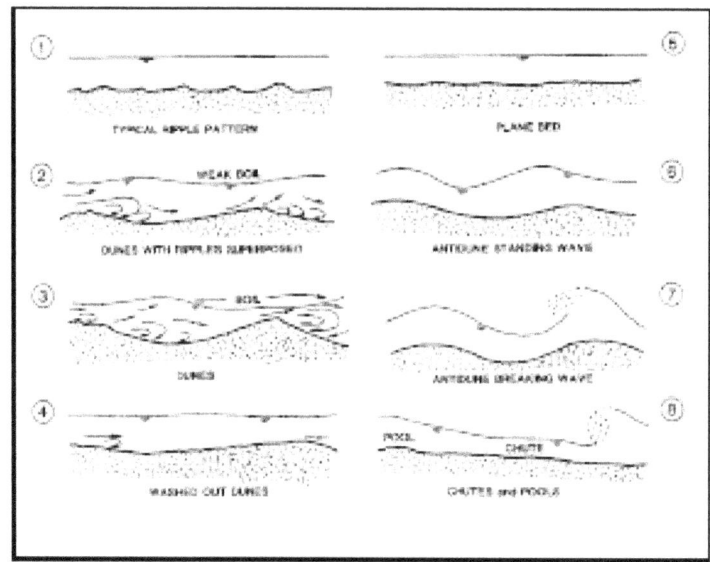

Figure 12. Forms of bed roughness in sand channels.

The resistance to flow for the different bedforms and coarser bed material will be given later in this section. For information on sediment transport and additional information on bedforms the reader is referred to (HIRE).[11]

At high flows, most sand-bed stream channels shift from a dune bed to a transition or a plane bed configuration. If the slope is steep antidune flow may occur. The resistance to flow is then decreased to one-half to one-third of that preceding the shift in bedform. The increase in velocity and corresponding decrease in depth may increase erosion and scour around bridge piers and abutments, increase the required size of riprap and decrease the required size of a drainage channel. However, if flow transitions to antidune flow, channel size may have to be increased in order to contain the waves that occur.

46

4.2.4. Coarse-Bed Material

At low flow, coarse alluvial bed material may not move, but at moderate or large flows, the material may become mobile. With the movement of coarse-bed material, large bars may form which will be residual at low flow. These bars can re-direct flow and cause bank erosion, scour holes, and clog drainage channels. Resistance to flow for coarse-bed material is caused by the grain roughness of the material and the form loss caused by the bars. However, coarse-bed material in drainage channels can have a beneficial effect by decreasing erosion by armoring of the bed. Information on armoring is given in HEC-18, HEC-20, and HIRE.[15,16,11] The determination of Manning's n for coarse-bed material is given later.

4.3. Steady Uniform Flow

In steady, uniform open-channel flow, there are no accelerations, streamlines are straight and parallel, and the pressure distribution is hydrostatic. The slope of the water surface S_w, the bed surface S_o, and the energy gradient S_f are equal (figure 6). It is the simplest flow condition to analyze. Steady uniform flow is an idealized concept for open-channel flow and is difficult to obtain even in laboratory flumes. For many applications, the flow is essentially steady and changes in width, depth, or direction (resulting in nonuniform flow) are so small that the flow can be considered uniform. In other cases, the changes occur over such a long distance the flow is a gradually varied flow.

The depth in steady uniform flow is called the normal depth and the symbol for it is given the subscript o as in Y_o. The velocity V is often given the same subscripts, i.e., V_o. Other variables of interest for steady uniform flow are (1) the discharge Q, (2) the velocity distribution v_y in the vertical, (3) the headloss H_L through the reach, and (4) the shear stress, both local and at the bed τ_o. All these variables are interrelated. In the following section, engineering equations will be given along with example problems for obtaining values for these variables.

4.3.1. Manning's Equation for Mean Velocity and Discharge

Water flows in a sloping drainage channel because of the force of gravity. The flow is resisted by the friction between the water and wetted surface of the channel. The quantity of water flowing (Q), the depth of flow (y), and the velocity of flow (V) depend upon the channel shape, roughness (n), and slope (S_o). Various equations have been devised to determine the velocity and discharge in open channels. A useful equation is the one that is named for Robert Manning, an Irish engineer. The metric form of Manning's equation for the velocity of flow in open channels is:

$$V = \frac{K_u}{n} R^{2/3} S^{1/2} \tag{13}$$

where:

    V = Mean velocity, m/s (ft/s)
    n = Manning's coefficient of channel roughness
    R = Hydraulic radius, m (ft)
    S = Energy slope, m/m (ft/ft)
    For steady uniform flow S = S_o.
    K_u = 1 (1.49)

Over many decades, a catalog of values of Manning's n has been assembled so that an engineer can estimate the appropriate value by knowing the general nature of the channel boundaries. An abbreviated list of Manning's roughness coefficients is given in table 12. Values for dredged and lined channels are given by references 17 and 11. For steeper streams, the reader is referred to reference 18. A pictorial guide for assisting with value selection is given by reference 13 in the Additional Reference listing.

An alternative approach for refining n value estimates consists of the selection of a base roughness value for a straight, uniform, and smooth channel in the materials involved and then adding values for the channel under consideration:

$$n = (n_0 + n_1 + n_2 + n_3 + n_4)m_5 \tag{14}$$

where:

n_0	=	Base value for straight uniform channels
n_1	=	Additive value due to cross-section irregularity
n_2	=	Additive value due to variations of the channel
n_3	=	Additive value due to obstructions
n_4	=	Additive value due to vegetation
m_5	=	Multiplication factor due to sinuosity

Detailed values of the coefficients are found in references 17, 19, 20, and 21. Typical values are given in tables B.2 or B.3. Reference 22 proposes a guide for selecting Manning's roughness coefficients for floodplains.

The roughness characteristics on the floodplain are complicated by the presence of vegetation, natural and artificial irregularities, buildings, undefined direction of flow, varying slopes, and other complexities. Resistance factors reflecting these effects must be selected largely on the basis of past experience with similar conditions. In general, resistance to flow is large on the floodplains. In some instances, conditions are further complicated by deposition of sediment and development of dunes and bars which affect resistance to flow and direction of flow.

The presence of ice affects channel roughness and resistance to flow in various ways. When an ice cover occurs, the open channel is more nearly comparable to a closed conduit. There is an added shear stress developed between the flowing water and ice cover. This surface shear is much larger than the normal shear stresses developed at the air-water interface. The ice-water interface is not always smooth. In many instances, the underside of the ice is deformed so that it resembles ripples or dunes observed on the bed of sand-bed channels. This may cause overall resistance to flow in the channel to be further increased. With total or partial ice cover, the drag of ice retards flow, decreasing the average velocity and increasing the depth.

The hydraulic radius, R, is a shape factor that depends only upon the channel dimensions and the depth of flow. It is computed by the equation:

$$R = \frac{A}{P} \tag{15}$$

where:

A	=	Cross-sectional area of the flowing water perpendicular to the direction of flow

P   =   Wetted perimeter or the length, of wetted contact between a stream of water and its containing channel, perpendicular to the direction of flow

The discharge Q is determined from the equation of continuity (see chapter 3). The equation is:

$$Q = AV \tag{16}$$

where:

Q   =   Discharge, m^3/s (ft^3/s)
A   =   Cross-sectional area, m^2 (ft^2)
V   =   Mean velocity, m/s (ft/s)

By combining equations 13 and 14, the Manning's equation can be used to compute discharge directly

$$Q = \frac{K_u}{n} A R^{2/3} S^{1/2} \tag{17}$$

In many computations, it is convenient to group the cross-sectional properties into a term called conveyance, K,

$$K = \frac{K_u}{n} A R^{2/3} \tag{18}$$

then

$$Q = K S^{1/2} \tag{19}$$

When a channel cross section is irregular in shape such as one with a relatively narrow deep main channel and wide shallow overbank area, the cross section must be subdivided and the flow computed separately for the main channel and overbank area. The same procedure is used when different parts of the cross section have different roughness coefficients. In computing the hydraulic radius of the subsections, the water depth common to the two adjacent subsections is not counted as wetted perimeter (see example problem 4.3).

Conveyance can be computed and a curve drawn for any channel cross section. The area and hydraulic radius are computed for various assumed depths and the corresponding value of K is computed from the equation. The values of conveyance are plotted against the depths of flow and a smooth curve connecting the plotted points is the conveyance curve. If the section was subdivided, the conveyance of each subsection (K_a, K_b,...K_n) is computed and the total conveyance of the channel is the sum of the conveyances of the subsections. Discharge can then be computed using equation 19. Example problem 4.3 illustrates a conveyance curve for a compound cross section. The concept of channel conveyance is useful when computing the distribution of overbank flood flows in the stream cross section and the distribution through the openings in a proposed stream crossing. The discharge through each opening can be assumed to have the same ratio to the total discharge as the ratio of conveyance of the opening bears to the total conveyance of the channel.

49

4.3.2. Aids in the Solution of Manning's Equation

Equations for the computation of Area A, wetted perimeter, P, and hydraulic radius, R, in rectangular and trapezoidal channels (figure 13) are:

$$A = By + Zy^2 \qquad\qquad (20)$$

$$P = B + 2y\sqrt{1 + Z^2} \qquad\qquad (21)$$

$$R = \frac{By + Zy^2}{B + 2y\sqrt{1 + Z^2}} \qquad\qquad (22)$$

The variables are defined in figure 13.

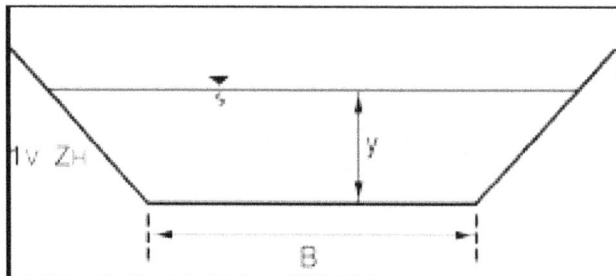

Figure 13. Trapezoidal channel.

4.3.3. Velocity Distribution

There are times in the design of highway drainage facilities that knowledge of the velocity distribution in the vertical is needed (e.g., the design of riprap for scour and erosion control). As a result of boundary roughness, the velocity varies vertically from some minimum value along the bed to a maximum value near the water surface (figure 14). In this section, the Einstein form of the Karman-Prandtl velocity distribution in the vertical and mean velocity equations will be given for steady uniform flow.[23] For their derivation the reader is referred to any standard fluid mechanics text or HIRE.[11]

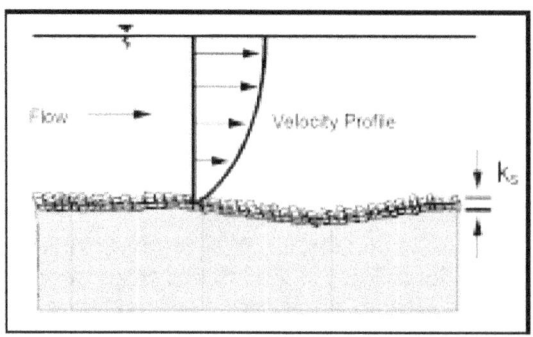

Figure 14. Schematic of vertical velocity profile.

EXAMPLE PROBLEM 4.1 (SI Units)

Given: Trapezoidal earth channel B = 2 m, sideslope 1V:2H, S = 0.003 m/m normal depth y = 0.5 m, n = 0.02.

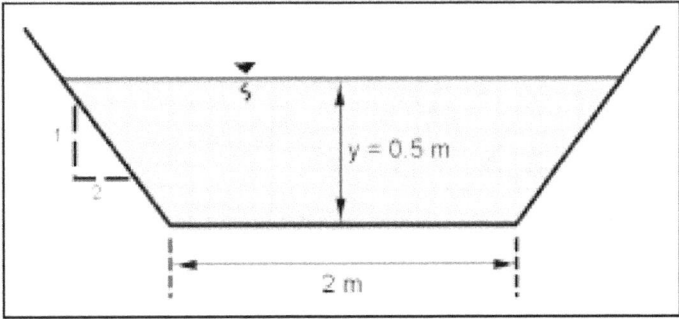

Find: Velocity V and discharge Q

Solution:

$$V = \frac{1}{n} R^{2/3} S^{1/2}$$

$$R = \frac{By + Zy^2}{B + 2y\sqrt{1+Z^2}} ; R = \frac{2(0.5) + 2(0.5)^2}{2 + 2(0.5)\sqrt{1+(2)^2}} = \frac{1.50}{4.24} = 0.35$$

$$V = \frac{1}{0.02}(0.35)^{2/3}(0.003)^{1/2} = 1.36 \text{ m/s}$$

$$Q = AV = (2(0.5) + 2(0.5)^2)1.36 = 2.0 \text{ m}^3/\text{s}$$

51

EXAMPLE PROBLEM 4.1 (English Units)

Given: Trapezoidal earth channel B = 6.5 ft, sideslope 1V:2H, S = 0.003 ft/ft normal depth y = 1.6 ft, n = 0.02.

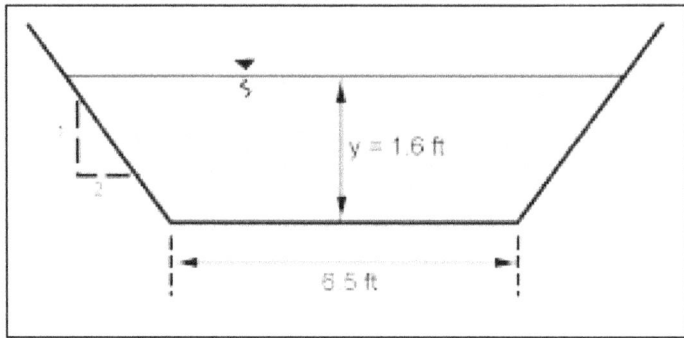

Find: Velocity V and discharge Q

Solution:

$$V = \frac{K_u}{n} R^{2/3} S^{1/2}$$

$$R = \frac{By + Zy^2}{B + 2y\sqrt{1 + Z^2}} \; ; R = \frac{(6.5)(1.6) + 2(1.6)^2}{6.5 + 2(1.6)\sqrt{1 + (2)^2}} = \frac{15.52}{13.66} = 1.14$$

$$V = \frac{1.49}{0.02} (1.14)^{2/3} (0.003)^{1/2} = 4.45 \; ft/s$$

$$Q = AV = (6.5)(1.6) + 2(1.6)^2) \, 4.45 = 69.06 \; ft^3/s$$

EXAMPLE PROBLEM 4.2 (SI Units)

Given: A concrete trapezoidal channel B = 1.5 m, sideslopes = 1V:2H, n = 0.013, slope = 0.002, Q = 3 m³/s

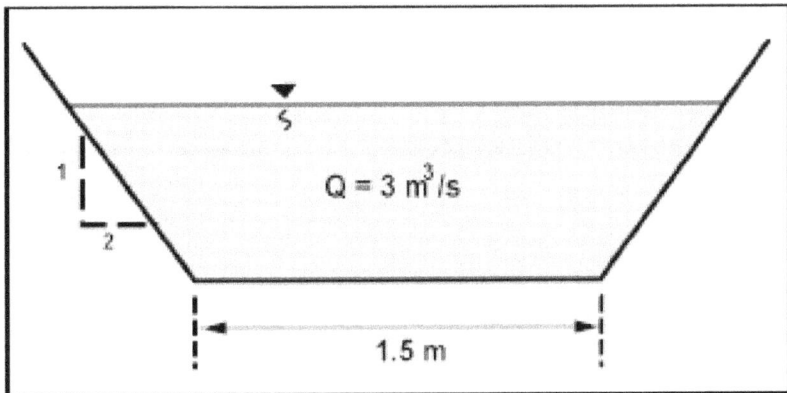

Find: Depth y and velocity v

Solution:

1. Use Manning's equation

$$Q = AV = A \frac{K_u}{n} R^{2/3} S^{1/2}$$

where K_u = 1 and relationships for A and R are

$$A = By + Zy^2 = 1.5y + 2y^2$$

$$R = \frac{By + Zy^2}{B + 2y\sqrt{1 + Z^2}} = \frac{1.5y + 2y^2}{1.5 + 4.47y}$$

substitute A and R into Manning's equation

$$Q = (1)\left(\frac{1.5y + 2y^2}{0.013}\right)\left(\frac{1.5y + 2y^2}{1.5 + 4.47y}\right)^{2/3} (0.002)^{1/2}$$

2. Trial and error solution for y to find a depth where Q = 3 m³/s

$$\text{Try } y = 0.70; \ Q = (1)\left(\frac{1.05 + 0.98}{0.013}\right)\left(\frac{1.05 + 0.98}{4.629}\right)^{2/3} (0.002)^{1/2} = 4.03$$

53

Since 4.03 > 3.0, the assumed value for y is too large. Try a smaller value such as 0.60.

$$\text{Try } y = 0.60; \; Q = (1)\left(\frac{0.9 + 0.72}{0.013}\right)\left(\frac{0.9 + 0.72}{4.18}\right)^{2/3} (0.002)^{1/2} = 2.96$$

since 2.96 = 3.0, the assumed value for y is okay.

Therefore, use y = 0.60 m and use continuity to find the velocity (V = Q/A)

$$V = \frac{3}{(0.9 + 0.72)} = 1.85 \text{m/s}$$

EXAMPLE PROBLEM 4.2 (English Units)

Given: A concrete trapezoidal channel B = 5.0 ft, sideslopes = 1V:2H, n = 0.013, slope = 0.002, Q = 105 ft³/s

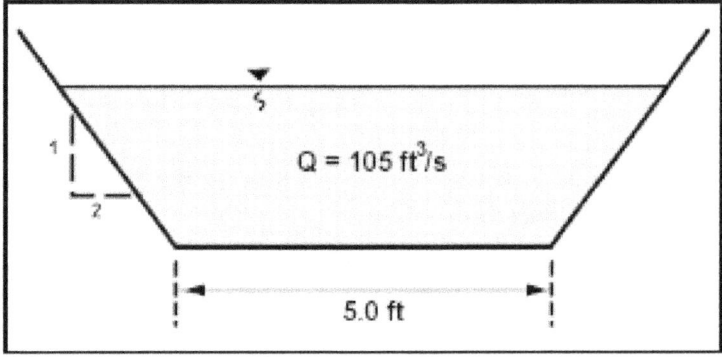

Find: Depth y and velocity v

Solution:

1. Use Manning's equation

$$Q = AV = A \frac{K_u}{n} R^{2/3} S^{1/2}$$

where K_u = 1.49 and relationships for A and R are

$$A = By + Zy^2 = 1.5y + 2y^2$$

$$R = \frac{By + Zy^2}{B + 2y\sqrt{1+Z^2}} = \frac{1.5y + 2y^2}{1.5 + 4.47y}$$

substitute A and R into Manning's equation

$$Q = (1.49)\left(\frac{1.5y + 2y^2}{0.013}\right)\left(\frac{1.5y + 2y^2}{1.5 + 4.47y}\right)^{2/3} (0.002)^{1/2}$$

2. Trial and error solution for y to find a depth where Q = 105 ft³/s

Try y = 2.5

$$Q = \frac{1.49}{(.013)}(1.5(2.5) + 2(2.5)^2)\left(\frac{1.5(2.5) + 2(2.5)^2}{1.5 + 4.47(2.5)}\right)^{2/3} (0.002)^{1/2} = 171.29 \, ft^3/s$$

55

Try y = 2.0 ft

$$Q = \frac{1.49}{(.013)}(1.5(2.0) + 2(2.0)^2)\left(\frac{1.5(2.0) + 2(2.0)^2}{1.5 + 4.47(2.0)}\right)^{2/3}(0.002)^{1/2} = 109.39\,\text{ft}^3/s$$

109 is close to 103, try 1.96 for y-

$$Q = \frac{1.49}{(.013)}(1.5(1.96) + 2(1.96)^2)\left(\frac{1.5(1.96) + 2(1.96)^2}{1.5 + 4.47(1.96)}\right)^{2/3}(0.002)^{1/2} = 105.11\,\text{ft}^3/s$$

Since 1.96 ft gave a Q of 105.11, the y of 1.96 is good Therefore use y = 1.96 and use continuity to find the velocity (V = Q/A)

$$V = \frac{105}{(1.5(1.96) + 2(1.96)^2)} = 6.01\,\text{ft}/\text{sec}$$

EXAMPLE PROBLEM 4.3 (SI Units)

Given: A compound channel as illustrated, with an n value of 0.03 and a longitudinal slope of 0.002 m/m.

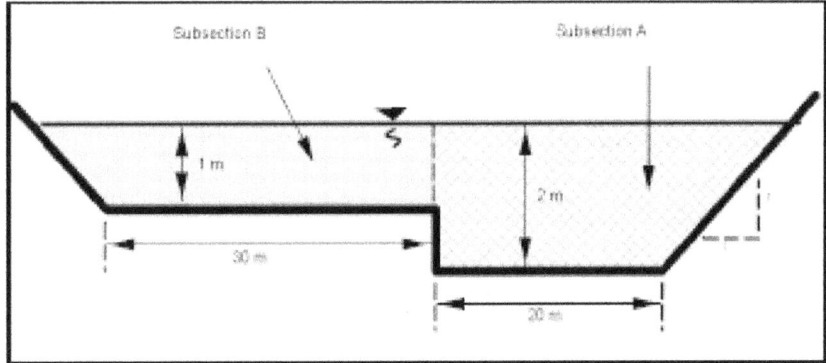

Find: Discharge Q

Solution:

1. Subsection A

$$A = (2)(20) + 1/2(2)(2) = 42 m^2$$

$$WP = 20 + (2)\sqrt{2} + 1 = 23.83 m$$

$$R = \frac{42}{23.83} = 1.76$$

$$V = \frac{1}{0.03}(1.76)^{2/3}(0.002)^{1/2} = 2.17 m/s$$

$$Q = 42(2.17) = 91.14 \ m^3/s$$

2. Subsection B

$$A = 30(1) + \frac{1}{2}(1)(1) = 30.5 m^2$$

$WP = 30 + 1\sqrt{2} = 31.41\,m$

$R = \dfrac{30.5}{31.41} = 0.97\,m$

$V = \dfrac{1}{0.03}(0.97)^{2/3}(0.002)^{1/2} = 1.46\,m/s$

$Q = 30.5\,(1.46) = 44.53\,m^3/s$

3. For the entire channel

$A = 42 + 30.5 = 72.5\,m^2$

$Q = 91.14 + 44.53 = 135.67\,m^3/s,\ say\ 136\,m^3/s$

4. If the channel had been considered as a whole without subdividing, the following results would have been obtained.

$A = 42 + 30.5 = 72.5\,m^2$

$WP = 23.82 + 31.41 = 55.23\,m$

$R = \dfrac{72.5}{55.23} = 1.31\,m$

$V = \dfrac{1}{0.03}(1.31)^{2/3}(0.002)^{1/2} = 1.78\,m/s$

$Q = 72.5\,(1.78) = 129.05\,m^3/s$

The discharge using the whole channel is considerably less than the discharge obtained by subdividing the channel. The difference would have been larger if the slope and/or the Manning's n would have been different. The problem also illustrates the difference in mean velocities in a shallow section as compared with the mean velocity of the main channel.

5. Plot the conveyance curve for this cross section and calculate the total discharge using equation 19.

• With a subdivided cross section, the conveyance at a given flow depth should be calculated for each subsection, and then added together to get the total conveyance at that depth. This calculation and a plot of the conveyance are illustrated below:

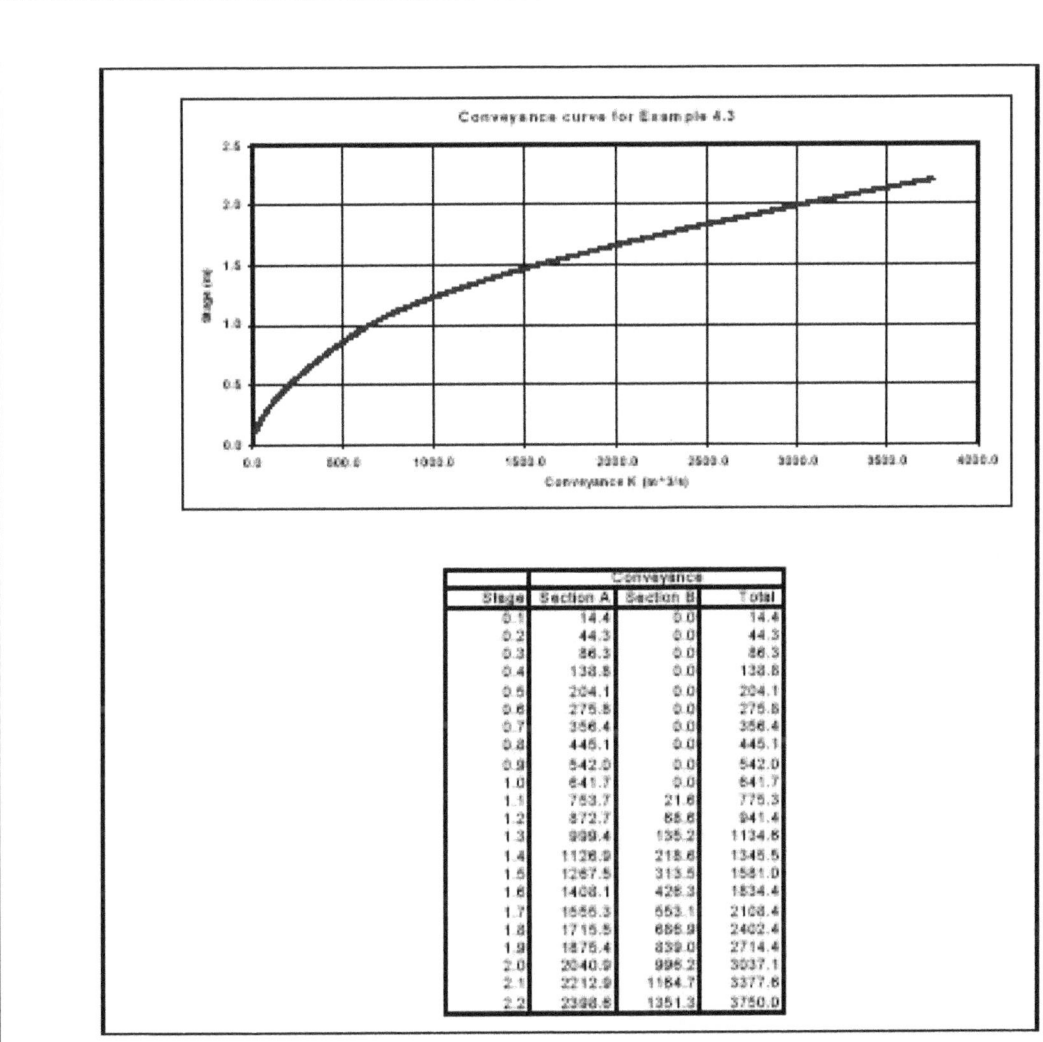

Conveyance curve for Example 4.3

Stage	Conveyance		
	Section A	Section B	Total
0.1	14.4	0.0	14.4
0.2	44.3	0.0	44.3
0.3	86.3	0.0	86.3
0.4	138.8	0.0	138.8
0.5	204.1	0.0	204.1
0.6	275.8	0.0	275.8
0.7	356.4	0.0	356.4
0.8	445.1	0.0	445.1
0.9	542.0	0.0	542.0
1.0	641.7	0.0	641.7
1.1	753.7	21.6	775.3
1.2	872.7	68.6	941.4
1.3	999.4	135.2	1134.6
1.4	1126.9	218.6	1345.5
1.5	1267.5	313.5	1581.0
1.6	1408.1	426.3	1834.4
1.7	1555.3	553.1	2108.4
1.8	1715.5	686.9	2402.4
1.9	1875.4	839.0	2714.4
2.0	2040.9	996.2	3037.1
2.1	2212.9	1164.7	3377.6
2.2	2398.6	1351.3	3750.0

- At a main channel flow depth of 2 m, the total conveyance is 3037.1. The total discharge is then (equation 19)

$$Q = K \, S^{1/2} = (3037.1)\,(0.002)^{1/2} = 135.8\,\text{m}^3\,/\,\text{s, say } 136\,\text{m}^3\,/\,\text{s}$$

This result matches the correct discharge value for a 2 m flow depth as calculated above in item 3.

EXAMPLE PROBLEM 4.3 (English Units)

Given: A compound channel as illustrated, with an n value of 0.03 and a longitudinal slope of 0.002 m/m.

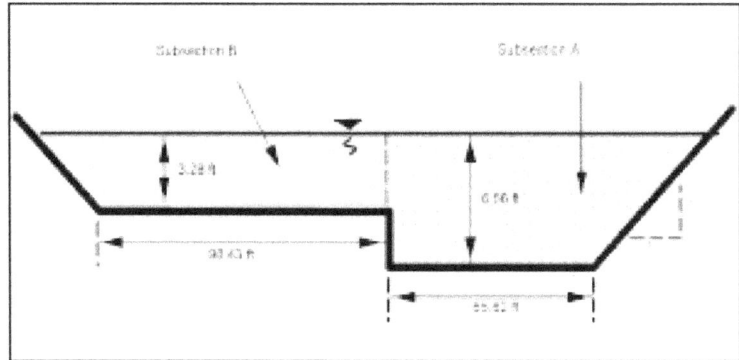

Find: Discharge Q

Solution:

1.   Subsection A

$A = (65.62)(6.56) + 1/2(6.56)(6.56) = 451.98 \text{ ft}^2$

$WP = 65.62 + (6.56)\sqrt{2} + 3.28 = 78.18 \text{ ft}$

$R = \dfrac{451.98}{78.18} = 5.78$

$V = \dfrac{1.49}{.03}(5.78)^{2/3}(.002)^{1/2} = 7.16 \text{ ft/s}$

$Q = (451.98)(7.16) = 3235 \text{ ft}^3/\text{s}$

2.   Subsection B

$A = (98.43)(3.28) + 1/2(3.28)(3.28) = 328.23 \text{ ft}^2$

$WP = 98.43 + 3.28 \sqrt{2} = 103.07 \, ft$

$R = \dfrac{328.23}{103.07} = 3.18$

$V = \dfrac{1.49}{.03}(3.18)^{2/3}(.002)^{1/2} = 4.81 \, ft/s$

$Q = (328.23)(4.81) = 1577 \, ft^3/s$

3.   For the entire channel

$A = 451.98 + 328.23 = 780.21 \, ft^2$

$Q = 3235 + 1577 = 4812 \, ft^3/s$

4.   If the channel had been considered as a whole without subdividing, the following results would have been obtained.

$A = 451.98 + 328.23 = 780.21 ft^2$

$WP = 78.18 + 103.07 = 181.25 \, ft$

$R = \dfrac{780.21}{181.25} = 4.30$

$V = \dfrac{1.49}{(.03)}(4.30)^{2/3}(0.002)^{1/2} = 5.88 \, ft/s$

$Q = (780.21)(5.88) = 4588 \, ft^3/s$

The discharge using the whole channel is considerably less than the discharge obtained by subdividing the channel. The difference would have been larger if the slope and/or the Manning's n would have been different. The problem also illustrates the difference in mean velocities in a shallow section as compared with the mean velocity of the main channel.

5.   Plot the conveyance curve for this cross section and calculate the total discharge using equation 19.

• With a subdivided cross section, the conveyance at a given flow depth should be calculated for each subsection, and then added together to get the total conveyance at that depth. This calculation and a plot of the conveyance are illustrated below:

61

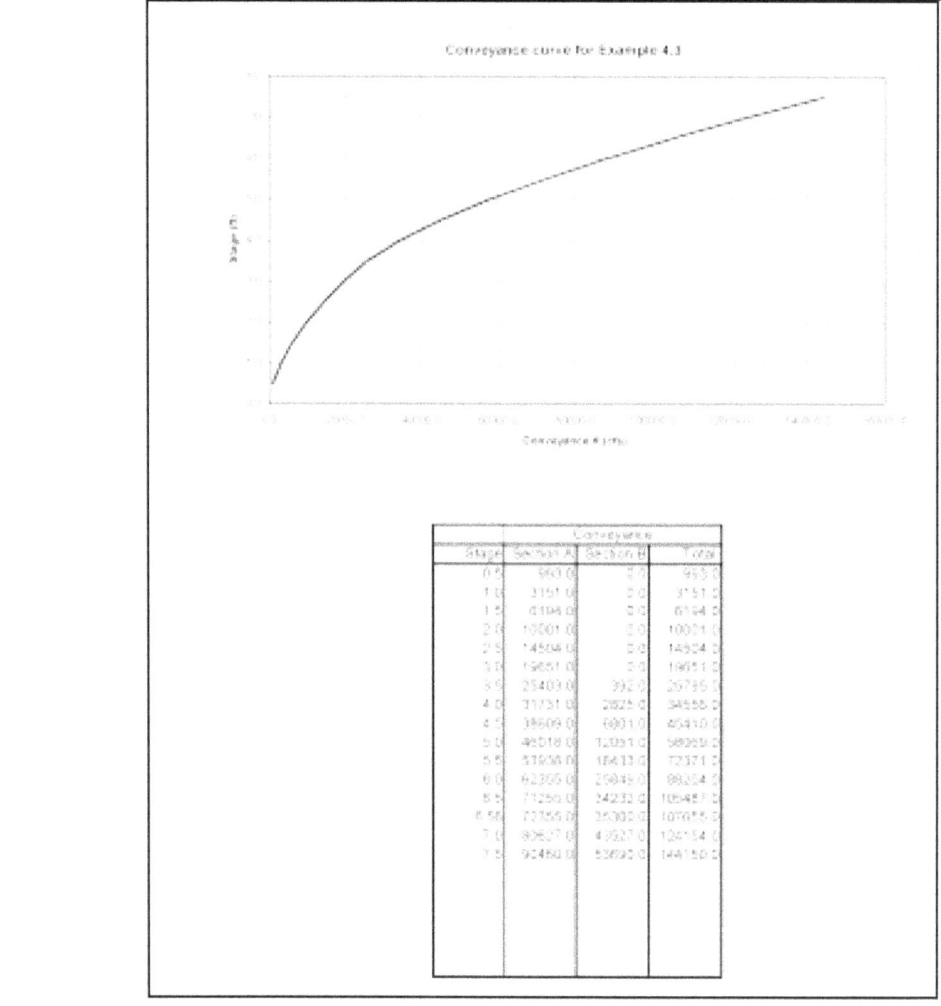

Conveyance curve for Example 4.3

	Conveyance		
Stage	Section A	Section B	Total
0.5	960.0	0.0	960.0
1.0	3151.0	0.0	3151.0
1.5	6194.0	0.0	6194.0
2.0	10001.0	0.0	10001.0
2.5	14504.0	0.0	14504.0
3.0	19651.0	0.0	19651.0
3.5	25403.0	392.0	25795.0
4.0	31731.0	2825.0	34556.0
4.5	38609.0	6801.0	45410.0
5.0	46018.0	12051.0	58069.0
5.5	53938.0	18433.0	72371.0
6.0	62305.0	25949.0	88254.0
6.5	71250.0	34232.0	105487.0
6.56	72355.0	35300.0	107655.0
7.0	80627.0	43527.0	124154.0
7.5	90460.0	53690.0	144150.0

- At a main channel flow depth of 6.56 ft, the total conveyance is 107,655. The total discharge is then (equation 19)

$$Q = K\,S^{1/2} = (107{,}655)(0.002)^{1/2} = 4814\ \text{ft}^3/\text{s}$$

This result matches closely to the 4,812 ft³/s calculated in item 3. The small difference is the result of rounding in the conveyance calculations.

The equations for velocity distribution (v) and mean velocity (V) can be written in the following dimensionless forms:

$$\frac{v}{V_*} = 5.75 \log \left(30.2 \frac{X y}{k_s}\right) \qquad (23)$$

and

$$\frac{V}{V_*} = 5.75 \log \left(12.27 \frac{X y_o}{k_s}\right) \qquad (24)$$

Note that any system of units can be used as long as y_o and k_s (and V, v and V_*) have the same units.

where:

X	=	Coefficient given in figure 15
k_s	=	Measure of the roughness height, k_s varies from the D_{84} size for pure sand bed channels, to 3.5 times D_{84} for graded coarse-bed streams; for practical application use 3.5 times D_{84}, m (ft)
y	=	Depth to specified location, m (ft)
v	=	Local mean velocity at depth y, m/s (ft/s)
y_o	=	Depth of flow, m (ft)
V	=	Depth-averaged velocity, m/s (ft/s)
V_*	=	Shear velocity, $(\tau_o/\rho)^{0.5}$, m/s (ft/s)
τ_o	=	Shear stress at the boundary, N/m² (lb/ft²)
δ'	=	Thickness of the viscous sublayer, 11.6 μ/V_*, m (ft)
μ	=	Dynamic viscosity of water, table 8, N-s/m² (lb-s/ft²)
ρ	=	Density of water, kg/m³ (lb-s²/ft⁴)

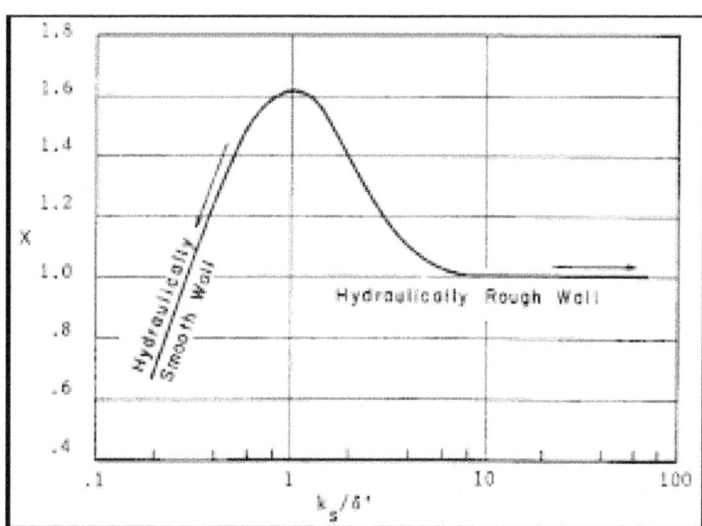

Figure 15. Einstein's multiplication factor X in the logarithmic velocity equations.[23]

63

4.3.4. Shear Stress

Shear stress is the force water exerts on the bed and bank of a channel as it flows over them. The following equations can be used to determine the shear stress on the boundary of the channel that results from the force of flowing water. For the derivations of these equations refer to fluid mechanics texts or HIRE.[11] The first equation (equation 25) is an exact equation, giving the average shear stress over the wetted perimeter. The next equations are semi-empirical and result from solving the Karman-Prandtl velocity equation.

$$\tau_0 = \gamma R S_0 \tag{25}$$

where:

τ_0	=	Average shear stress on the wetted perimeter, N/m^2 (lb/ft^2)
γ	=	The unit weight of water, N/m^3 (lb/ft^3)
R	=	Hydraulic radius, m (ft)
S_0	=	Slope of the channel, m/m (ft/ft). In gradually varied flow the slope is of the energy grade line, $S_0 = S_f$

$$\tau_0 = \frac{\rho (v_1 - v_2)^2}{[5.75 \log(\frac{y_1}{y_2})]^2} \tag{26}$$

$$\tau_0 = \frac{\rho V^2}{[5.75 \log(12.27 \frac{y_0}{k_s})]^2} \tag{27}$$

where τ_0 is the shear stress at a point in the flow, N/m^2 (lb/ft^2), v_1 and v_2 are point velocities in the vertical at y_1 and y_2, respectively; V is the mean velocity in the vertical with a depth of y_0; and the other terms have been defined previously.

4.3.5. Froude Number and Relationship to Subcritical, Critical, and Supercritical Flow

An extremely important dimensionless parameter in open-channel flow is the Froude Number, defined as the ratio of the inertia forces to the gravity forces. It is normally expressed as

$$Fr = \frac{V}{\sqrt{gy}} \tag{28}$$

where:

Fr	=	Froude Number
V	=	Velocity of flow, m/s (ft/s)
g	=	Acceleration of gravity, m/s^2 (ft/s^2)
y	=	Depth of flow, m (ft)

EXAMPLE PROBLEM 4.4 (SI Units)

Determine the shear stress along the wetted perimeter of a trapezoidal channel. Also determine the shear stress on a particle along the bottom of the same channel.

Given: Trapezoidal channel as illustrated with S_o = .005, γ = 9800 N/m^3, V = 1.8 m/s, D_{84} = 0.15m

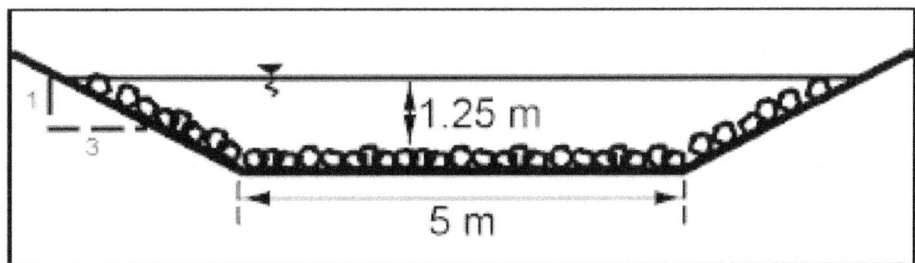

Find:

(1) τ_o along wetted perimeter
(2) τ_o along bed

Solution:

(1)   the shear stress along the wetted perimeter is given by

$$\tau_o = \gamma RS$$

where

$$R = \frac{A}{P}$$

$$A = 5 \times 1.25 + 3\,(1.25)^2 = 10.938\ m^2$$

$$P = 5 + 2 \times 1.25\,\sqrt{10} = 12.905\ m$$

$$R = \frac{10.938}{12.905} = .848\ m$$

$$\tau_o = 9800\ N/m^3\,(0.85\ m)\,(0.005\ m/m) = 41.6\ N/m^2$$

(2)   the shear stress along the bottom at a point is

$$\tau_o = \frac{\rho \, V^2}{[5.75 \log (12.27 \frac{y_o}{k_s})]^2}$$

$$\tau_o = \frac{1000 \, kg/m^3 \, (1.8 \, m/s)^2}{[5.75 \log (12.27 \frac{1.25 \, m}{(3.5 \times 0.15 \, m)})]^2} = 45.6 \, N/m^2$$

EXAMPLE PROBLEM 4.4 (English Units)

Determine the shear stress along the wetted perimeter of a trapezoidal channel. Also determine the shear stress on a particle along the bottom of the same channel.

Given: Trapezoidal channel as illustrated with S_o = .005, γ = 62.4 lb/ft^3, V = 5.9 ft/s,
$\quad\quad$ D_{84} = 0.49 ft

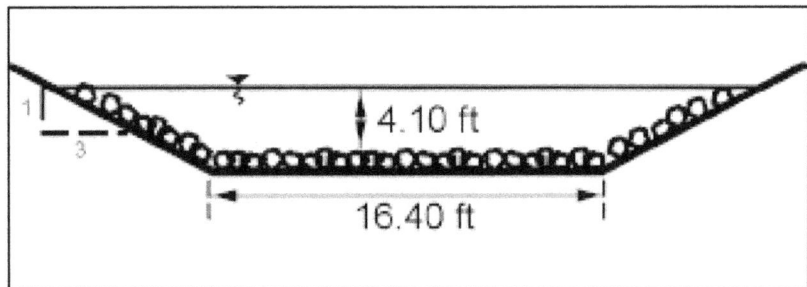

Find:

(1)$\quad$ τ_o along wetted perimeter
(2)$\quad$ τ_o along bed

Solution:

(1)$\quad$ the shear stress along the wetted perimeter is given by

$$\tau_o = \gamma R S$$

where

$$R = \frac{A}{P}$$

$$A = 16.4 \times 4.10 + 3(4.10)^2 = 117.67 \text{ ft}^2$$

$$P = 16.4 + 2 \times 4.10 \sqrt{10} = 42.33 \text{ ft}$$

$$R = \frac{117.67}{42.33} = 2.78 \text{ ft}$$

$$\tau_o = 62.4 \text{ lb/ft}^3 \ (2.78 \text{ ft})(.005) \text{ ft/ft} = .87 \text{ lb/ft}^2$$

(2) the shear stress along the bottom at a point is

$$\tau_o = \frac{\rho\,V^2}{[5.75 \log (12.27 \frac{y_o}{k_s})]^2}$$

$$\rho = 1.94\ \text{lb} - \text{s}^2 / \text{ft}^4$$

$$\tau_o = \frac{1.94\,\text{lb} - \text{s}^2 / \text{ft}^4\,(5.9\,\text{ft}/\text{s})^2}{\left[5.75 \log (12.27) \dfrac{4.10\,\text{ft}}{(3.5 \times .49\,\text{ft})} \right]^2} = 0.95\ \text{lb} / \text{ft}^2$$

V and y can be the mean velocity and depth in a channel or the velocity and depth in the vertical. If the former are used, then the Froude Number is for the average flow conditions in the channel. If the latter are used, then it is the Froude Number for that vertical at a specific location in the cross section. The Froude Number uniquely describes the flow pattern in open-channel flow. For example, in alluvial channel flow with sand-bed material, ripples and dunes only form when the Froude Number is less than 1.0 (subcritical flow); whereas, antidunes only form when the Froude Number is greater than 1.0. Plane bed formation is independent of the Froude Number. The Froude Number is the scaling parameter that is used in modeling open channel flow structures in the laboratory.

When the Froude Number is 1.0, the flow is critical; values of the Froude Number greater than 1.0 indicate supercritical or rapid flow and smaller than 1.0 indicate subcritical or tranquil flow. The velocity and depth at critical flow are called the critical velocity and critical depth. The channel slope which produces critical depth and critical velocity is the critical slope. The change from supercritical to subcritical flow is often abrupt (particularly if the Froude Number is larger than 2.0) resulting in a phenomenon known as the hydraulic jump.

Critical depth and velocity for a particular discharge are only dependent on channel size and shape and are independent of channel slope and roughness. Critical slope depends upon the channel roughness, channel geometry, and discharge. For a given critical depth and velocity, the critical slope for a particular roughness can be computed by Manning's equation.

Supercritical flow is difficult to control because abrupt changes in alignment or in cross section produce waves which travel downstream, alternating from side to side, sometimes causing the water to overtop the channel sides. Changes in channel shape, slope, alignment, or roughness cannot be reflected upstream. In supercritical flow, the control of the flow is located upstream. Supercritical flow is common in steep flumes, channels, and mountain streams.

Subcritical flow is relatively easy to control for flows with Froude Numbers less than 0.8. Changes in channel shape, slope, alignment, and roughness affect the flow for small distances upstream. The control in subcritical flow is located downstream. Subcritical flow is common in channels, flumes and streams located in the plains regions and valleys where slopes are relatively flat.

Critical depth is important in hydraulic analysis because it is always a hydraulic control. The flow must pass through critical depth in going from subcritical flow to supercritical or going from supercritical flow to subcritical. Although, in the latter case a hydraulic jump usually occurs. Typical locations of critical depth are:

1.   At abrupt changes in slope when a flat (subcritical) slope is sharply increased to a steep (supercritical) slope.

2.   At channel constrictions such as a culvert entrance, flume transitions, etc. under some conditions.

3.   At the unsubmerged outlet of a culvert or flume on a subcritical slope, discharging into a wide channel, steep slope channel (supercritical), or with a free fall at the outlet.

4.   At the crest of an overflow dam, weir, or embankment.

5.   At bridge constrictions where the bridge chokes the flow.

The location and magnitude of critical depth and the determination of critical slope for a cross section of a given shape, size, and roughness are important in channel design and analysis. The equations for determining the critical depth are provided in the discussion of specific discharge and specific energy in steady rapidly varied flow (section 4.6).

4.4.   Unsteady Flow

Unsteady flows of interest to the highway drainage engineer or designer are:

1.   Waves resulting from disturbances of the water surface by wind and boats.

2.   Waves resulting from the surface instability that exists for flows with Froude Numbers close to 1.0.

3.   Waves resulting from flow disturbance due to change in direction of flow with Froude Numbers greater than about 2.0.

4.   Surges or bores resulting from sudden increase or decrease in the flow by opening or closing of gates or the movement of tides on coastal streams.

5.   Standing waves and antidunes that occur in alluvial channel flow.

6.   Flood waves resulting from the progressive movement downstream of stream runoff or gradual release from reservoirs.

Waves are an important consideration in bridge hydraulics when designing slope protection of embankments and dikes, and channel improvements. In the following paragraphs, only the basic one-dimensional analysis of waves and surges is presented. Other aspects of waves are presented in other sections.

69

4.4.1. Gravity Waves

For shallow water waves (long waves - figure 16) where the normal depth (y_o) is small in comparison to the wave length, the basic equation for the celerity (velocity of the wave relative to the velocity of the flow) is given by:

$$c = \sqrt{gy_o} \qquad (29)$$

Note that the celerity of a shallow water wave of small amplitude is the same as the denominator of the Froude Number.

$$Fr = \frac{V}{\sqrt{gy_o}} \qquad (30)$$

As explained in the discussion of the Froude Number (section 4.3.5), when Fr < 1 (subcritical or tranquil), a small amplitude wave moves upstream. When Fr > 1 (supercritical or rapid flow), a small amplitude wave moves downstream and when Fr = 1 (critical flow), a small amplitude wave is stationary. The fact that waves or surges cannot move upstream when the Froude Number is equal to or greater than 1.0 is important to remember when determining when the stage-discharge relation at a cross section can be affected by downstream conditions.

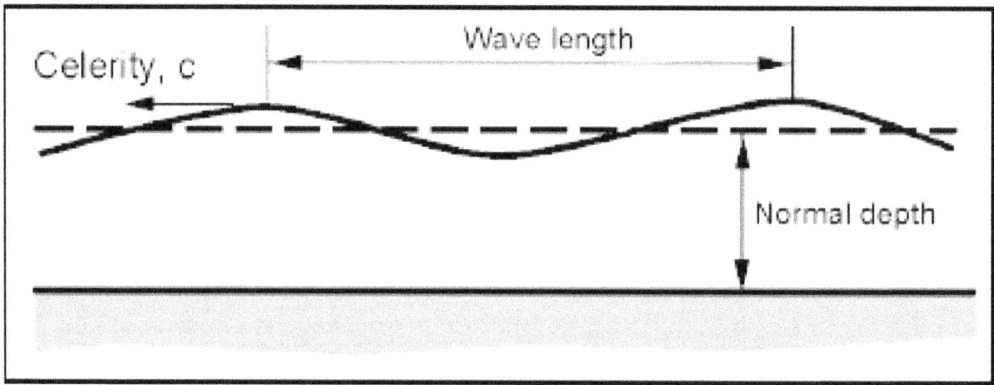

Figure 16. Definition sketch for small amplitude waves.

4.4.2. Surges

A surge is a rapid increase in the depth of flow (figure 17a). A surge may result from the sudden release of water from a dam or an incoming tide. The lifting of a gate in a channel not only causes a positive surge to move downstream, it also causes a negative surge to move upstream (figure 17b). As it moves upstream, a negative surge quickly flattens out. See HIRE for more detail and the basic surge equation.[11]

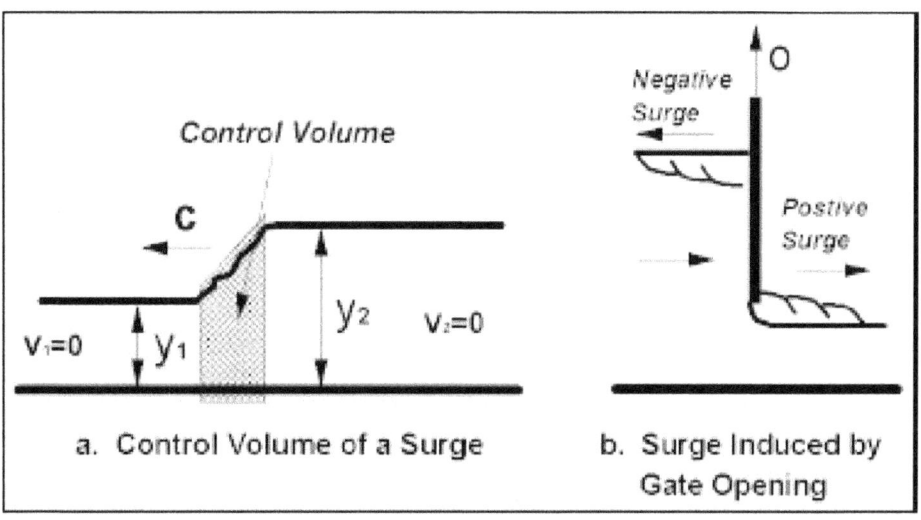

Figure 17. Sketch of positive and negative surges.

4.5. Steady Nonuniform Flow

Steady nonuniform flow occurs when the quantity of water (discharge) remains constant, but the depth of flow, velocity, or cross section changes from section to section. From the continuity equation, the relation of all cross sections will be:

$$Q = A_1 V_1 = A_2 V_2 = A_n V_n \tag{31}$$

Velocity in steady nonuniform flow can be computed using Manning's equation if the change in velocity from section to section is small so that the effect of acceleration is small.

The hydraulic design engineer needs a knowledge of nonuniform flow in order to determine the behavior of the flowing water when changes in channel resistance, size, cross section, shape, or slope occur. Typical examples might include determining water surface elevation changes in a channel of constant slope that goes through a short transition from a concrete trapezoidal cross section (with a low Manning's n) to a larger grass-lined trapezoidal cross section (a high Manning's n), or a stream with constant slope and Manning's n that is a long distance upstream of a culvert that constricts the flow.

These two situations define two basic cases of steady nonuniform flow. The first case is for relatively short distances (a few meters (feet) to several hundred meters (feet)) where accelerations are more important than friction. This case is called STEADY RAPIDLY VARIED FLOW. The effect of friction, if it is important, is taken into account by subdividing the distance into shorter segments and using Manning's equation along these shorter segments. The second case is for long distances (hundreds to thousands of meters (feet)), where friction losses are more important than accelerations. This case is called GRADUALLY VARIED FLOW. The method of analysis and equations for these two cases will be given in the next two sections.

71

4.6.   Steady Rapidly Varied Flow

4.6.1. Introduction

Steady flow through relatively short transitions where the flow is uniform before and after the transition can be analyzed using the energy equation. Energy loss due to friction may be neglected, at least as a first approximation. Refinement of the analysis can be made in a second step by including friction loss. For example, the water surface elevation through a transition is determined using the energy equation and then modified by determining the friction loss effects on velocity and depth in short subsections through the transition. However, energy losses resulting from flow separation cannot be neglected, and transitions where separation may occur need special treatment which may include model studies. Contracting flows (converging streamlines) are less susceptible to separation than expanding flows. Also, any time a transition changes velocity and depth such that the Froude Number approaches unity, problems such as waves, blockage, or choking of the flow may occur. If the approaching flow is supercritical, a hydraulic jump may result. Transitions for supercritical flow are discussed in a later section.

Transitions are used to contract or expand a channel width (figure 18a), to increase or decrease bottom elevation (figure 18b), or to change both the width and bottom elevation. The analysis or design of transitions is aided by the use of the depth of flow and velocity head terms in the energy equation (see chapter 3). The sum of the two terms is called the specific energy or specific head, H, and defined as

$$H = \frac{V^2}{2g} + y = \frac{q^2}{2gy^2} + y \qquad\qquad (32)$$

where:

H	=	Specific energy, m (ft)
q	=	Unit discharge, defined as the discharge per unit width m^3/s/m (ft^3/s/ft)in a rectangular channel
g	=	Acceleration of gravity, 9.81 m/s^2 (32.2 ft/s^2)
y	=	Depth of flow, m (ft)

The specific energy, H, is the height of the total energy above the channel bed.

The relationship between the three terms in the specific energy equation, q, y, and H, are evaluated by considering q constant and determining the relationship between H and y (specific energy diagram) or considering H constant and determining the relationship between q and y (specific discharge diagram). These diagrams for a given discharge or energy are then used in the design or analysis of transitions or flow through bridges. They are explained in the next two sections.

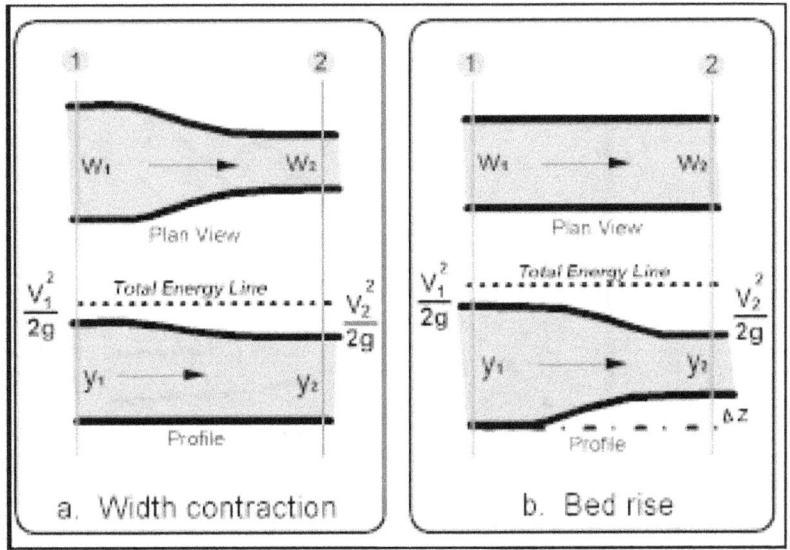

Figure 18. Transitions in open-channel flow (subcritical flow).

4.6.2. Specific Energy Diagram and Evaluation of Critical Depth

For a given q, equation 32 can be solved for various values of H and y. When y is plotted as a function of H, figure 19 is obtained. There are two possible depths called alternate depths for any H larger than a specific minimum. Thus, for specific energy larger than the minimum, the flow may have a large depth with small velocity or small depth with large velocity. Flow for a given unit discharge q cannot occur with specific energy less than the minimum. The single depth of flow at the minimum specific energy is called the critical depth, y_c, and the corresponding velocity, the critical velocity, $V_c = q/y_c$. The relation for y_c and V_c for a given q (for a rectangular channel) is

$$y_c = 3\sqrt{\frac{q^2}{g}} = 2\frac{V_c^2}{2g} \tag{33}$$

Note that for critical flow:

$$\frac{V_c}{\sqrt{gy_c}} = 1 = Fr \tag{34}$$

and

$$H_{min} = \frac{V_c^2}{2g} + y_c = \frac{3}{2}y_c \tag{35}$$

73

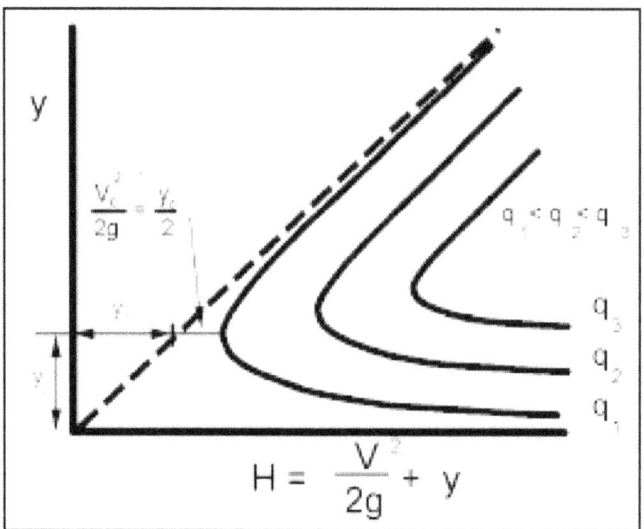

Figure 19. Specific head diagram.

Thus, flow at minimum specific energy has a Froude Number equal to 1. Flows with velocities larger than critical (Fr > 1) are called rapid or supercritical and flow with velocities smaller than critical (Fr < 1) are called tranquil or subcritical. These flow conditions are illustrated in figure 20 where a rise in the bed causes a decrease in depth when the flow is tranquil and an increase in depth when the flow is rapid. Furthermore there is a maximum rise in the bed for a given H_1 where the given rate of flow is physically possible. If the rise in the bed is increased beyond Δz_{max} for H_{min} then the approaching flow depth y_1 would have to increase (increasing H) or the flow would have to be decreased. Thus, for a given flow in a channel, a rise in the bed level can occur up to a Δz_{max} without causing backwater.

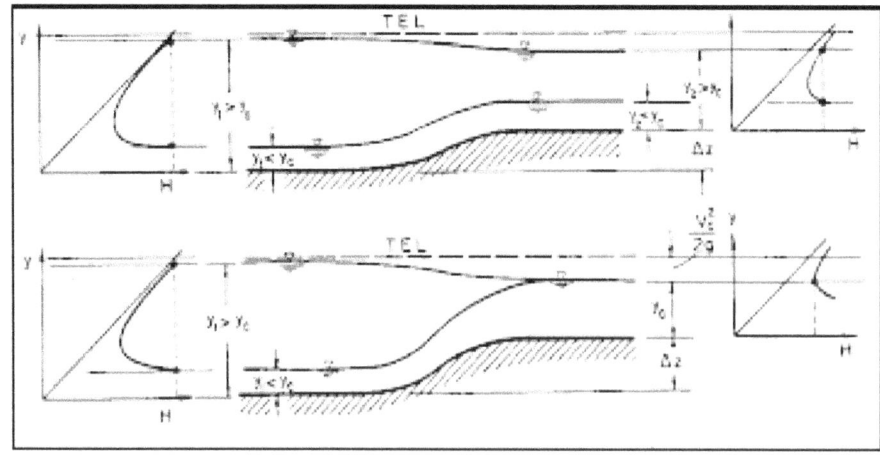

Figure 20. Changes in water surface resulting from an increase in bed elevation.

74

Distinguishing between the types of flow and how the water surface reacts with changes in cross section is important in channel design; thus, the location of critical depth and the determination of critical slope for a cross section of given shape, size, and roughness becomes necessary. Equations for direct solution of the critical depth are available for several prismatic shapes; however, some of these equations were not derived for use in the metric system.[12]

For any channel section, regular or irregular, critical depth may be found by a trial-and-error solution of the following equation:

$$\frac{A_c^3}{T_c} = \frac{Q^2}{g} \tag{36}$$

where A_c and T_c and the area and topwidth at critical flow. An expression for the critical velocity (V_c) of any cross section at critical flow conditions is:

$$V_c = \sqrt{gy_c} \tag{37}$$

where: $y_c = A_c / T_c$ (38)

Uniform flow within about 10 percent of the critical depth is unstable and should be avoided in design. The reason for unstable flow can be seen by referring to the specific head diagram (figure 19). As the flow approaches the critical depth from either limb of the curve, a very small change in energy is required for the depth to abruptly change to the alternate depth on the opposite limb of the specific head curve. If the unstable flow region cannot be avoided in design, the least favorable type of flow should be assumed for the design.

4.6.3. Specific Discharge Diagram

Equation 32 can be rearranged to determine q, the unit discharge, as a function of H, the specific energy, and y, the depth of flow.

$$q = y\sqrt{2g\,(H = y)} \tag{39}$$

For a constant H, q can be solved as a function of y and the specific discharge diagram will result (figure 21).

For any discharge smaller than a specific maximum q for the given H, two depths of flow are possible. The depth at maximum q for a given specific energy H, is the critical depth, y_c and the velocity is the critical velocity, V_c.

$$y_c = \frac{2}{3}H = 2\frac{V_c^2}{2g} \tag{40}$$

and

$$\frac{V_c}{\sqrt{gy_c}} = 1 = Fr \tag{41}$$

75

EXAMPLE PROBLEM 4.5 (SI Units)

Given: Determine the critical depth in a trapezoidal shaped swale with z = 1, given a discharge of 9.2 m³/s and a bottom width, B = 6 m. Also, determine the critical velocity.

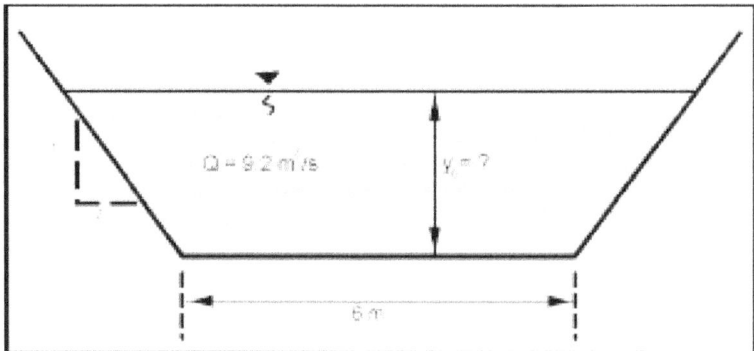

Find: Critical depth

Velocity at critical depth

Solution:

For a Q of 9.2 m³/s

$$\frac{A_c^3}{T_c} = \frac{Q^2}{9}$$

$$\frac{A_c^3}{T_c} = \frac{(9.2)^2}{9.81} = 8.63$$

For a trapezoidal channel area substituting A = y (B + zy) and T = B + 2zy gives

$$8.63 = \frac{[y(6+y)]^3}{6+2y}$$

A trial and error solution yields y = 0.6 m.

$$V_c = \sqrt{g\, y_c} \qquad y_c = \frac{A}{T} = \frac{0.6\,\text{m}\,(6\,\text{m}+0.6)}{(6\,\text{m}+1.2\,\text{m})} = 0.55\,\text{m}$$

$$V_c = \sqrt{9.81(.55)} = 2.3\,\text{m/s}$$

EXAMPLE PROBLEM 4.5 (English Units)

Given: Determine the critical depth in a trapezoidal shaped swale with z = 1, given a discharge of 325 ft³/s and a bottom width, B = 20 ft. Also, determine the critical velocity.

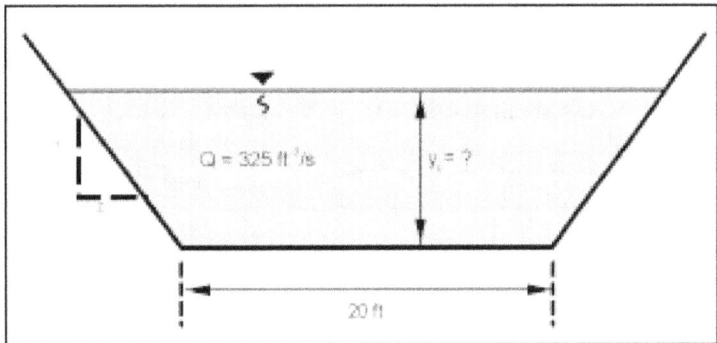

Find: Critical depth

Velocity at critical depth

Solution:

For a Q of 325 ft³/s

$$\frac{A_c^3}{T_c} = \frac{Q^2}{g}$$

$$\frac{A_c^3}{T_c} = \frac{(325 \text{ ft}^3/\text{s})^2}{32.2 \text{ ft}/\text{s}^2} = 3280.28$$

For a trapezoidal channel area substituting A = y (B + zy) and T = B + 2zy gives

$$3280 = \frac{[y(20 + y)]^3}{20 + 2y}$$

A = 1.95 (20 + (1) 1.95) = 42.80 ft²
T = 20 + 2 (1) (1.95) = 23.90 ft

A trial and error solution yields y = 1.95 ft.

$$V_c = \sqrt{g\,y_c} \qquad y_c = \frac{A}{T} = \frac{42.80)}{23.9} = 1.79 \text{ ft}$$

$$V_c = \sqrt{32.2(1.79)} = 7.59 \text{ ft}/\text{s}$$

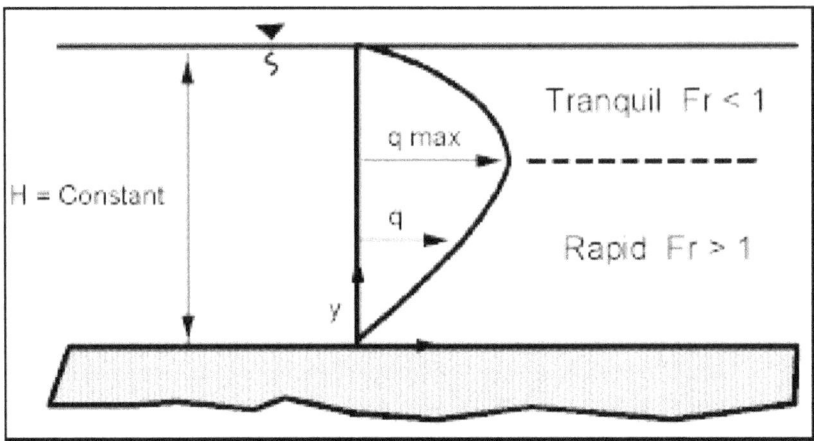

Figure 21. Specific discharge diagram.

Thus, for maximum discharge at constant H, the Froude Number is 1.0, and the flow is critical. The relation between y_c, V_c, H and q_{max} for a constant H is

$$y_c = \frac{2}{3}H = 3\sqrt{\frac{q_{max}^2}{g}} = 2\frac{V_c^2}{2g} \qquad (42)$$

Flow conditions for constant specific energy for a width contraction are illustrated in figure 22 assuming no geometrical effects such as eccentricity, skew, piers, scour, and expansion. The contraction causes a decrease in flow depth when the flow is tranquil and an increase when the flow is rapid. The maximum possible contraction without causing backwater effects occurs when the Froude Number is 1.0, the discharge per unit of width q is a maximum, and y_c is 2/3 H. A further decrease in width will cause backwater. That is, an increase in depth upstream will occur to produce a larger specific energy and increase y_c in order to get the flow through the decreased width.

The flow in figure 22 can go from point A to C and then either back to D or down to E depending on the downstream boundary conditions. An increase in slope of the bed downstream from C and no separation would allow the flow to follow the line A to C to E. Similarly, the flow can go from B to C and back to E or up to D depending on boundary conditions. Figure 22 is drawn with the side boundary forming a smooth streamline. If the contraction was due to bridge abutments, the upstream flow would follow a natural streamline to the point of maximum constriction, but then downstream, the flow would probably separate. Tranquil approach flow could follow line A to C but the downstream flow probably would not follow either line C to D or C to E, but would have an undulating hydraulic jump. There would be interaction of the flow in the separation zone and considerable energy would be lost. If the slope downstream of the abutments was the same as upstream, then the flow could not be sustained with this amount of energy loss. Backwater would occur, increasing the depth in the constriction and upstream, until the flow could go through the constriction and establish uniform flow downstream.

78

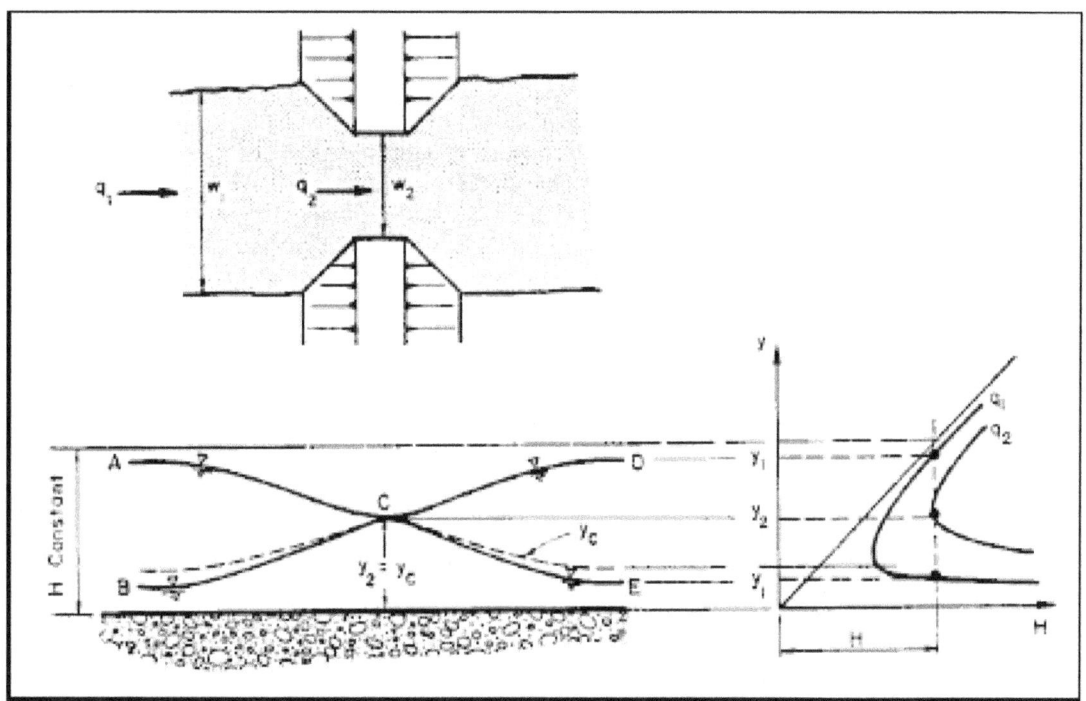

Figure 22. Change in water surface elevation resulting from a change in width.

4.6.4. Hydraulic Jump

A hydraulic jump will occur when the flow velocity V_1 is rapid or supercritical and the slope is decreased to a slope for subcritical flow, or an obstruction such as an energy dissipater is placed in the flow. The supercritical depth is changed to a subcritical depth, called the sequent depth. Depending on the magnitude of the Froude Number, a considerable amount of energy is changed to heat. The larger the Froude Number, the more energy that is lost. The existence of a jump assumes adequate tailwater conditions exist. Many engineers/designers assume that a jump will always occur when a change from a steep grade to a flat grade is encountered, such as near the outlet end of a culvert (i.e., a broken-back culvert). The jump will only occur with adequate downstream tailwater to maintain the sequent depth just below the culvert grade break. Without adequate tailwater, the jump will be swept downstream out of the culvert, causing a potentially large scour hole at the culvert outlet.

The relation between the supercritical depth and the sequent depth for a rectangular flat channel is

$$\frac{y_2}{y_1} = \frac{1}{2}\{(1 + 8\ Fr_1^2)^{1/2} - 1\} \tag{43}$$

The corresponding energy loss in a hydraulic jump is the difference between the two specific energies. It can be shown that this headloss is

$$\Delta H = h_L = \frac{(y_2 - y_1)^3}{4\ y_1\ y_2} \tag{44}$$

79

Equation 44 has been experimentally verified along with the dependence of the jump length L_j and energy dissipation (headloss h_L) on the Froude Number of the approaching flow (Fr_1). The results of these experiments are given in figure 23.

When the Froude Number for rapid flow is less than 2.0, an undulating jump with large surface waves is produced. The waves are propagated for a considerable distance downstream. In addition, when the Froude Number of the approaching flow is less than 3.0, the energy dissipation of the jump is not large and jets of high velocity flow can exist for some distance downstream of the jump. These waves and jets can cause erosion a considerable distance downstream of the jump. For larger values of the Froude Number, the rate of energy dissipation in the jump is very large and figure 23 is recommended.

4.6.5. Subcritical Flow in Bends

When subcritical flow goes around a bend, the water surface is elevated on the outside of the bend and lowered on the inside of the bend (figure 24). The approximate difference in elevation (ΔZ) between the water surface along the sides of the curved channel can be found by the following equations.

$$\Delta Z = Z_0 - Z_i = \frac{V^2}{gr_c}(r_0 - r_i)$$ (45)

where:

Z	=	Elevation of the water surface, m (ft)
V	=	Average velocity in the channel, m/s (ft/s)
g	=	Acceleration of gravity, 9.81m/s^2 (32.2 ft/s^2)
r_c	=	Radius of curvature to the centerline of the channel, m (ft)
r_0	=	Radius of curvature to the outside flow line around the bend, m (ft)
r_i	=	Radius of curvature to the inside flow line around the bend, m (ft)

In channel design, superelevation is accounted for by adding $\Delta Z/2$ to the normal depth to define the maximum water surface depth at the outside of the bend.

This equation gives values of ΔZ somewhat lower than will occur in natural channels because of the assumption of uniform velocity and uniform curvature, but the computed value will be generally less than 10 percent in error. An equation from HIRE (below) is more accurate.[11] The difference in superelevation obtained by using the two equations is small, and in alluvial channels, the resulting erosion of the concave bank and deposition on the convex bank leads to further error in computing superelevation. Therefore, it is recommended that the first equation (equation 45) be used to compute superelevation in alluvial channels. For lined canals with strong curvature and large velocities, superelevation should be computed using

$$\Delta Z = \frac{V_{max}^2}{2g}\left\{ 2 - \left(\frac{r_i}{r_c}\right)^2 - \left(\frac{r_c}{r_0}\right)^2 \right\}$$ (46)

EXAMPLE PROBLEM 4.6 (SI Units)

Given: A hydraulic jump occurs in a 5-m wide rectangular channel at a flow depth of 0.5m. Determine the downstream water surface elevation needed to cause the jump. Also calculate the headloss due to the jump. Given Q = 20 m^3/s.

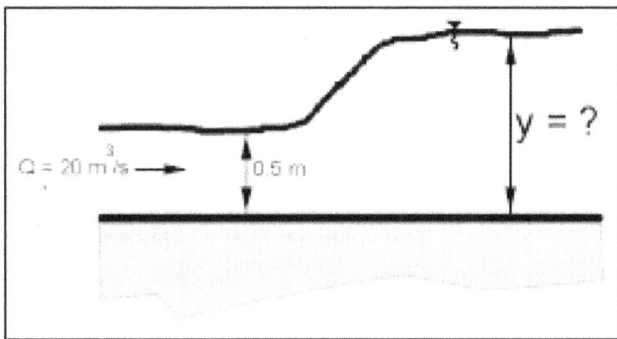

Find:

(1)   Determine the required downstream WSEL to initiate a jump
(2)   Determine the headloss across the jump

Solution:

(1)   Using the formula for a hydraulic jump, find y_2.

From continuity $V_1 = \dfrac{Q}{A_1} = \dfrac{20}{5(0.5)} = 8\,m/s$

$y_2 = \dfrac{y_1}{2}\left(\sqrt{1+8Fr_1^2} - 1\right); \; Fr_1 = \dfrac{V_1}{\sqrt{gy_1}} = \dfrac{8}{\sqrt{9.81(0.5)}} = 3.6$

$y_2 = 2.3\,m$

(2)   Find the headloss, h_L, across the jump

$h_L = \dfrac{(y_2 - y_1)^3}{4(y_2)(y_1)}$

$h_L = \dfrac{(2.3 - 0.5)^3}{4\,(2.3)\,(0.5)} = 1.27\,m$

81

EXAMPLE PROBLEM 4.6 (English Units)

Given: A hydraulic jump occurs in a 16.4 ft wide rectangular channel at a flow depth of 1.64 ft. Determine the downstream water surface elevation needed to cause the jump. Also calculate the headloss due to the jump. Given Q = 700 ft³/s.

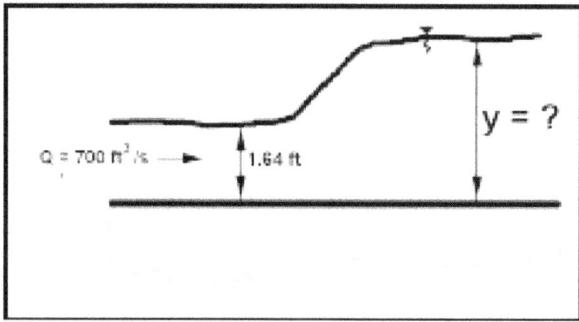

Find:

(1)   Determine the required downstream WSEL to initiate a jump
(2)   Determine the headloss across the jump

Solution:

(1)   Using the formula for a hydraulic jump, find y_2.

$$\text{From continuity } V_1 = \frac{Q}{A_1} = \frac{700}{(1.64)(16.4)} = 26.03 \text{ ft/s}$$

$$y_2 = \frac{y_1}{2}\left(\sqrt{1+8Fr_1^2} - 1\right); \; Fr_1 = \frac{V_1}{\sqrt{gy_1}} = \frac{26.03}{\sqrt{32.2(1.64)}} = 3.58$$

$$y_2 = \frac{1.64}{2}\left(\sqrt{1+8(3.58)^2} - 1\right) = 7.52 \text{ ft}$$

(2)   Find the headloss, h_L, across the jump

$$h_L = \frac{(y_2 - y_1)^3}{4(y_2)(y_1)}$$

$$h_L = \frac{(7.52 - 1.64)^3}{4(7.52)(1.64)} = 4.12 \text{ ft}$$

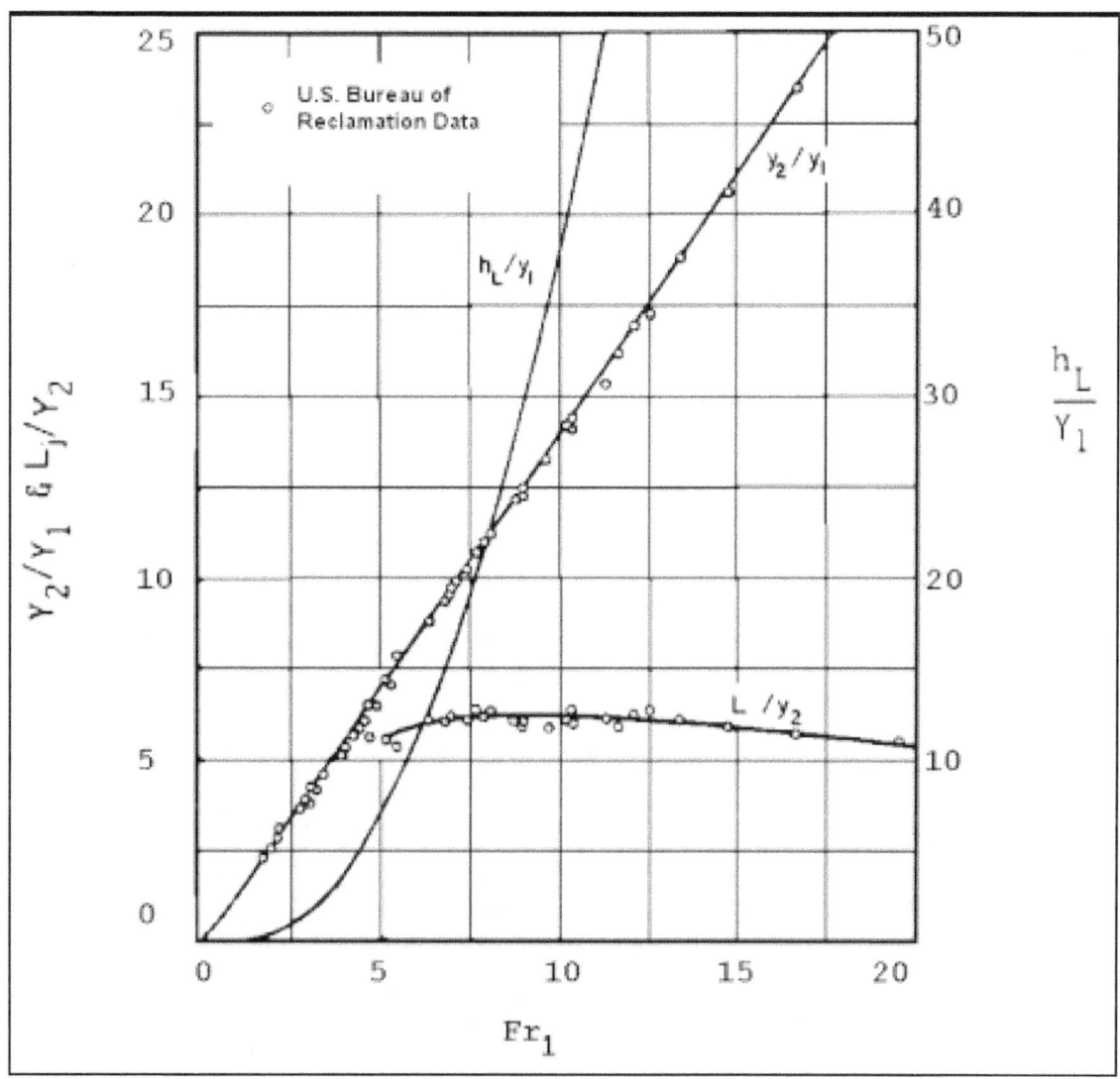

Figure 23. Hydraulic jump characteristics as a function of the upstream Froude Number.

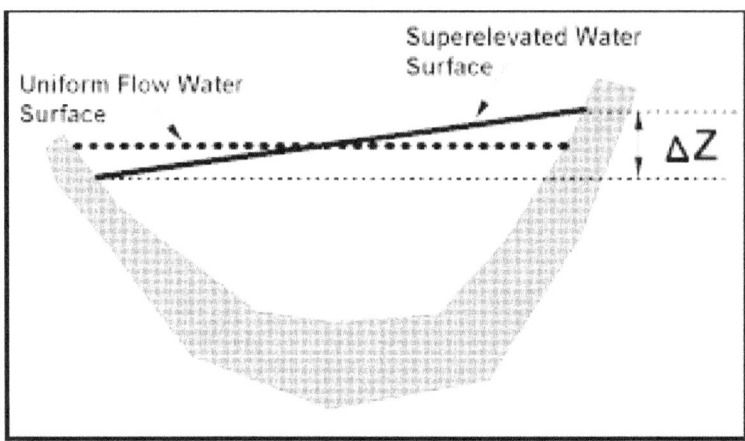

Figure 24. Superelevation of water surface in a bend.

Other problems introduced by curved alignment of channel in subcritical flow include spiral flow, changes in velocity distribution, and increased friction losses within the curved channel as contrasted with the straight channel. Flow-around bends are discussed in references 11, 17, and 24.

4.6.6. Supercritical Flow in Bends

Changes in alignment of supercritical flow are difficult to make. The water traveling at supercritical velocities around bends builds up waves which may "climb out" of the channel. The waves that are set up may continue downstream for a long distance. Also, sharp changes in alignment may set up a hydraulic jump with the flow overtopping the banks. Changes in alignment, whenever possible, should be made near the upper end of the section before the supercritical velocity has developed. If a change in alignment is necessary in a channel carrying supercritical flow, the channel should be rectangular in cross section, and preferably enclosed. On small chutes, experiments have shown that an angular variation (α) of rectangular flow boundaries (expansion) should not exceed that produced by the equation:

$$\tan \alpha = \frac{1}{3Fr} \tag{47}$$

Changes in alignment of open channels can and should be designed to reduce the wave action, resulting from the change in direction in flow (see reference 11). Often designs involving supercritical flow should be model tested to develop the best design, or even a design that will work.

84

EXAMPLE PROBLEM 4.7 (SI Units)

Given: During high runoff, a 2.0 m deep mountain stream flows near bank full with a normal depth and velocity of 1.8m and 3.4 m/s, respectively. At a sharp bend r_o = 12 m, r_c = 10 m, r_i = 8 m. Will flow overtop the bend?

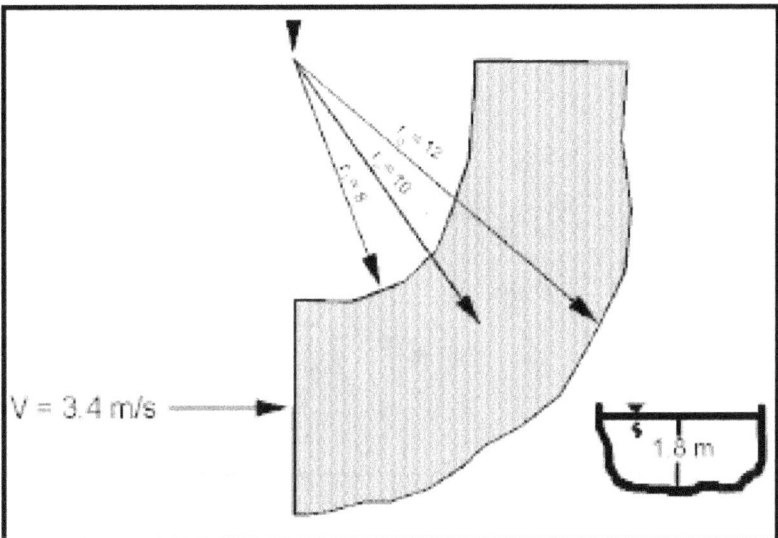

Find: ΔZ

Solution: Use the superelevation formula

$$\Delta Z = \frac{V^2}{gr_c}(r_o - r_i) = \frac{3.4^2}{9.81(10)}(12 - 8) = 0.47 \text{ m}$$

The water surface raises approximately (0.47/2) m on the outside of the bend and lowers by that same amount on the inside of the bend. The maximum flow depth in the bend will be

$$y_{outside} = 1.8 + \frac{0.47}{2} = 2.04 \text{ m}$$

which is greater than the channel depth (2.0 m) and overtopping will occur.

EXAMPLE PROBLEM 4.7 (English Units)

Given: During high runoff, a 6.56 ft deep mountain stream flows near bank full with a normal depth and velocity of 5.91 ft and 11.15 ft/s, respectively. At a sharp bend $r_o = 39.37$ ft, $r_c = 32.81$ ft, $r_i = 26.25$ ft. Will flow overtop the bend?

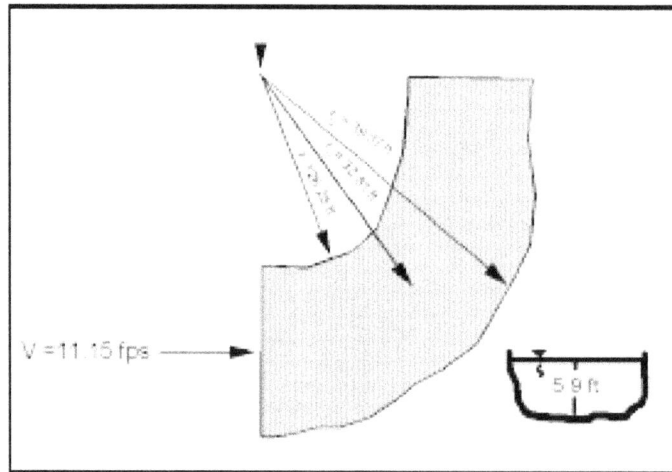

Find: ΔZ

Solution: Use the superelevation formula

$$\Delta Z = \frac{V^2}{gr_c}(r_o - r_i) = \frac{(11.15)^2}{32.2(32.81)}(39.37 - 26.25) = 1.54 \text{ ft}$$

The water surface raises approximately (1.54/2) ft on the outside of the bend and lowers by that same amount on the inside of the bend. The maximum flow depth in the bend will be

$$y_{outside} = 5.9 + \frac{1.54}{2} = 6.67 \text{ ft}$$

which is greater than the channel depth (6.56 ft) and overtopping will occur.

4.7   Gradually Varied Flow

4.7.1. Introduction

In this section, a second type of steady nonuniform flow is considered, gradually varied flow. In gradually varied flow, changes in depth and velocity take place slowly over large distances, resistance to flow dominates, and acceleration forces are neglected. Analysis of gradually varied flow involves: (1) the determination of the general characteristics of the water surface; and (2) the elevation of the water surface or depth of flow.

In gradually varied flow, the actual flow depth, y, is either larger or smaller than the normal depth, y_o, and either larger or smaller than the critical depth, y_c. The water surface profiles, which are often called backwater curves, depend on the magnitude of the actual depth of flow, y, in relation to the normal depth, y_o, and the critical depth, y_c. Normal depth, y_o, is the depth of flow that would exist for steady-uniform flow as determined using Manning's velocity equation, and the critical depth is the depth of flow when the Froude Number equals 1.0. Reasons for the depth being different than the normal depth are changes in slope of the bed, changes in cross section, obstruction to flow, and imbalances between gravitational forces accelerating the flow and shear forces retarding the flow. In working with gradually varied flow, the first step is to determine what type of water surface profile would exist. The second step is to perform the numerical computations.

4.7.2. Types of Water Surface Profiles

The types of water surface profiles are obtained by analyzing the change of the various terms in the gradually varied flow equation.[11]

$$\frac{dy}{dx} = S_o \left\{ \frac{1 - \left(\frac{n}{n_o}\right)^2 \left(\frac{y_o}{y}\right)^{\frac{10}{3}}}{1 - \left(\frac{y_c}{y}\right)^3} \right\} \tag{48}$$

The slope of the water surface dy/dx depends on the slope of the bed S_o, the ratio of the normal depth y_o to the actual depth y and the ratio of the critical depth y_c to the actual depth y. The difference between flow resistance for steady uniform flow n_o to flow resistance for steady nonuniform flow n is small and the ratio is taken as 1.0. With n = n_o, there are 12 types of water surface profiles. The 12 types are subdivided into 5 classes which depend on the bed slope. These are illustrated in figure 25 and summarized in table 1.

When y → y_c, the assumption that acceleration forces can be neglected no longer holds. Equation 48 indicates that dy/dx is perpendicular to the bed slope when y → y_c. For locations close to the cross-section where flow is critical, a distance from 3 to 20 m, (10 to 65 ft) curvilinear flow analysis and experimentation must be used to determine the actual values of y. When analyzing long distances, 30 to 100 m or longer, (100 to 300 ft or longer) one can assume qualitatively that y reaches y_c. In general, when the flow is rapid (Fr > 1), the flow cannot become tranquil or subcritical without a hydraulic jump occurring. In contrast, subcritical flow can become rapid, or supercritical, (cross the critical depth line). This is illustrated in figure 26.

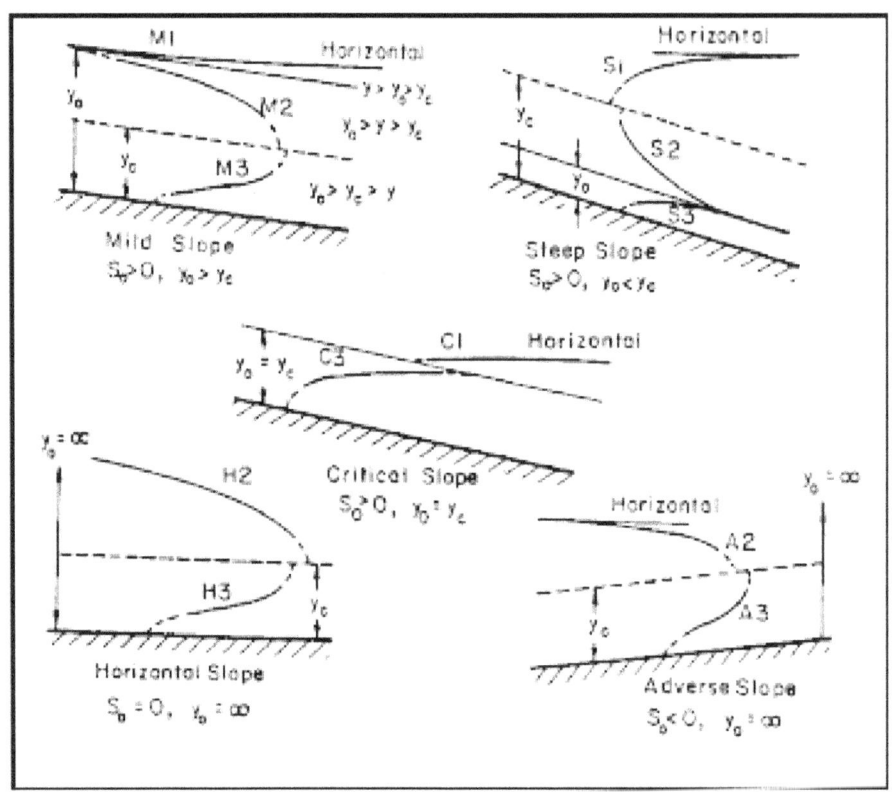

Figure 25. Classification of water surface profiles.

When there is a change in cross section or slope or an obstruction to the flow, the qualitative analysis of the flow profile depends on locating the control points, determining the type of water surface profile upstream and downstream of the control points, and then sketching these profiles. It must be remembered that when flow is supercritical (Fr > 1), the control depth is upstream and the water surface profile analysis proceeds in the downstream direction. When flow is subcritical (Fr < 1), the control depth is downstream and the computations must proceed upstream. The water surface profiles that result from a change in slope of the bed are illustrated in figure 26.

To determine if the hydraulic jump occurs on the steep or mild slope, calculate the sequent depth (y_2) for the y_1 depth using the hydraulic jump equation. If y_2 from the hydraulic jump is larger than the normal depth y_0 from Manning's equation on the mild slope, then there will be an M3 curve on the mild slope until the y_2 equals the depth that corresponds to the initial depth needed for the jump to occur. If y_2 is smaller than the depth that would balance with the downstream depth, the jump will occur on the steep slope and an S1 curve will occur to connect with the normal depth at the control section.

88

Table 1. Characteristics of Water Surface Profiles.				
Class	Bed Slope	Depth	Type	Classification
Mild	$S_c > 0$	$y > y_o > y_c$	1	M1
Mild	$S_c > 0$	$y_o > y > y_c$	2	M2
Mild	$S_o > 0$	$y_o > y_c > y$	3	M3
Critical	$S_o > 0$	$y > y_o = y_c$	1	C1
Critical	$S_c > 0$	$y < y_o = y_c$	3	C3
Steep	$S_o > 0$	$y > y_c > y_o$	1	S1
Steep	$S_o > 0$	$y_c > y > y_o$	2	S2
Steep	$S_o > 0$	$y_c > y_o > y$	3	S3
Horizontal	$S_o = 0$	$y > y_c$	2	H2
Horizontal	$S_o = 0$	$y_c > y$	3	H3
Adverse	$S_o < 0$	$y > y_c$	2	A2
Adverse	$S_o < 0$	$y_c > y$	3	A3

Note:

1. With a _type 1 curve_ (M1, S1, C1), the actual depth of flow y is greater than both the normal depth y_o and the critical depth y_c. Because flow is tranquil, control of the flow is downstream.

2. With a _type 2 curve_ (M2, S2, A2, H2), the actual depth y is between the normal depth y_o and the critical depth y_c. The flow is tranquil for M2, A2 and H2 and thus the control is downstream. Flow is rapid for S2 and the control is upstream.

3. With a _type 3 curve_ (M3, S3, C3, A3, H3), the actual depth y is smaller than both the normal depth y_o and the critical depth y_c. Because the flow is rapid control is upstream.

4. For a _mild slope_, S_o is smaller than S_c and $y_o > y_c$.

5. For a _steep slope_, S_o is larger than S_c and $y_o < y_c$.

6. For a _critical slope_, S_o equals S_c and $y_o = y_c$.

7. For an _adverse slope_, S_o is negative.

8. For a _horizontal slope_, S_o equals zero.

9. The case where $y \to y_c$ is of special interest because the denominator in equation 48 approaches zero.

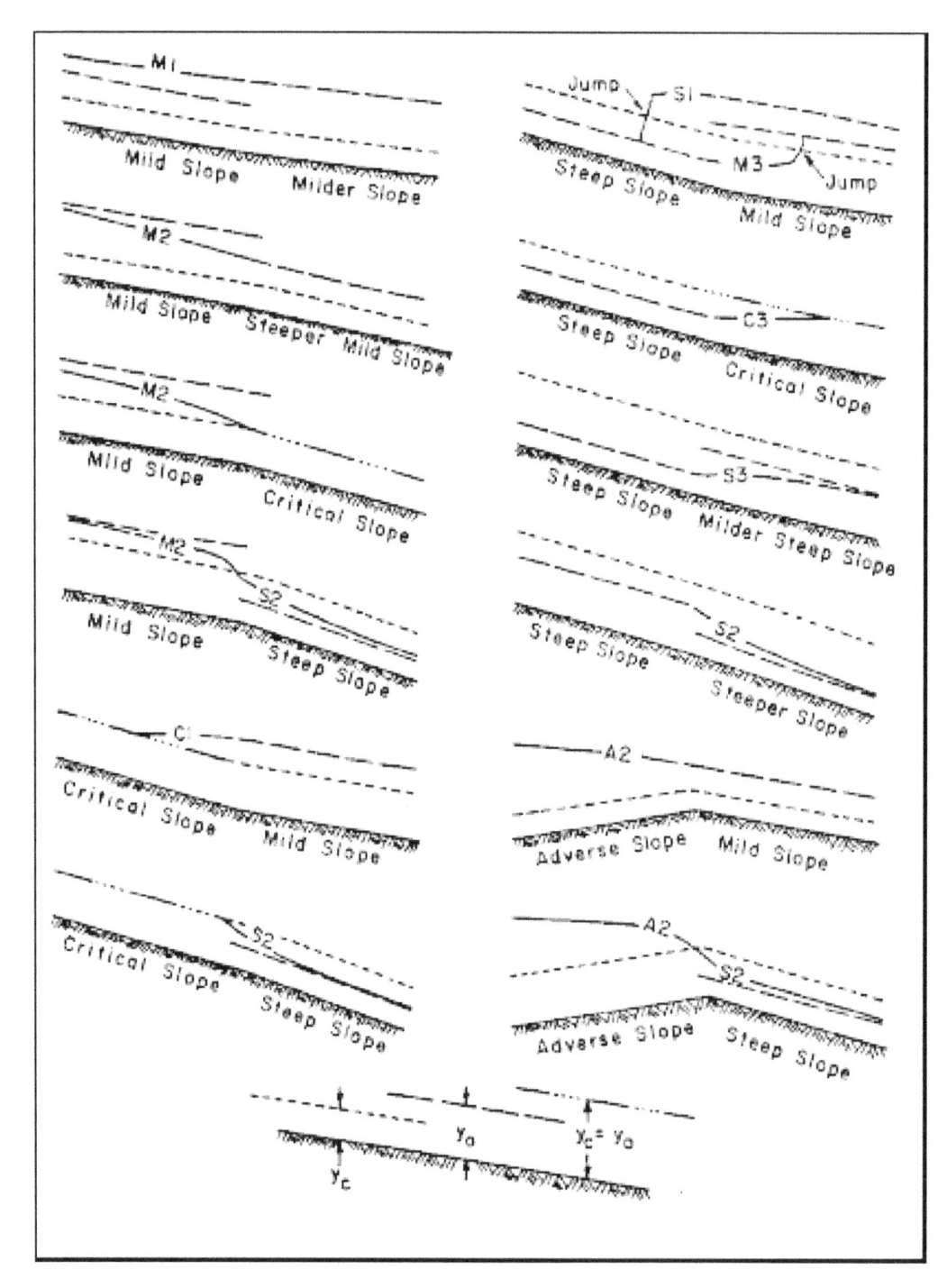

Figure 26. Examples of water surface profiles.

EXAMPLE PROBLEM 4.8 (SI Units)

Given: A 5 m wide, rectangular channel goes from a very steep grade to a mild slope. The design discharge is 24.8 m³/s and the normal depth and velocity on the steep slope were calculated to be 0.33 m and 15 m/s, respectively. On the mild slope, the normal depth and velocity were calculated to be 2.96 m and 1.68 m/s, respectively. Determine the type of flow occurring in both channels. If a hydraulic jump occurs, evaluate the depth downstream of the hydraulic jump, the location of the jump, and the water surface profile classification.

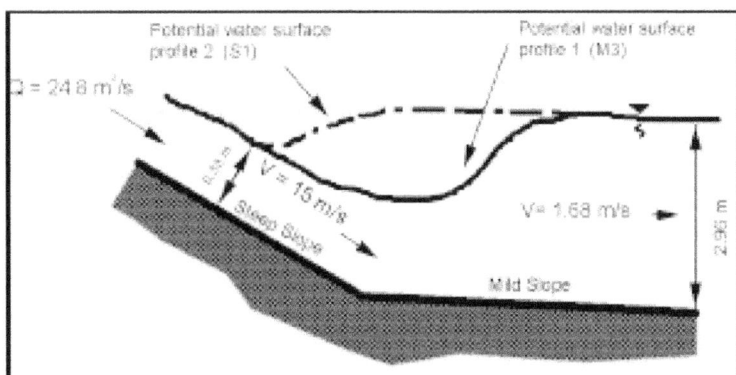

Find:

1.   Find the critical depth, y_c, on the steep slope.

$$y_c = \left(\frac{q^2}{g} \right)^{\frac{1}{3}} \qquad q = Q/B \text{ where } B \text{ - channel width}$$

$$y_c = \left[\frac{(Q/B)^2}{g} \right]^{\frac{1}{3}} = \left[\frac{(24.8/5)^2}{9.81} \right]^{\frac{1}{3}} = 1.36 \, m$$

On the steep slope, the normal depth is 0.33 m. Since $y < y_c$, supercritical flow occurs on the steep slope. Note that the unit discharge, q, is the same for the mild slope and hence, y_c, is the same for the steep and mild slope sections. On the mild slope, the normal depth is 2.96 m. Since $y > y_c$, subcritical flow occurs on the mild slope. Therefore, a hydraulic jump should occur.

2.   Next, determine if the jump will occur on the steep slope or on the mild slope.

To determine if the hydraulic jump occurs on the steep or mild slope, calculate the sequent depth (y_2) for the steep slope y_1 depth using the hydraulic jump equation. If y_2 from the hydraulic jump is larger than the normal depth y_0 from Manning's equation on the mild slope, then there will be an M3 curve on the mild slope until the y_2 equals the critical depth. If y_2 is smaller than y_0 on the mild slope, then the jump may occur on the steep slope and an S1 curve will occur to connect with the normal depth at the control section.

$$y_2 = \frac{y_1}{2}\left(\sqrt{1+8\,Fr_1^2} - 1\right)$$

$$Fr = \frac{V}{\sqrt{gy}} = \frac{15}{\sqrt{9.81(0.33)}} = 8.34$$

$$y_2 = \frac{0.33}{2}\left(\sqrt{1+8\,(8.34)^2} - 1\right) = 3.73\text{ m}$$

Compare parameters and determine the type of water surface classification.

Since the sequent depth (y_2) is greater than the mild slope normal depth (i.e., 3.73 > 2.96), the mild slope channel will have an M3 curve until the hydraulic jump occurs.

EXAMPLE PROBLEM 4.8 (English Units)

Given: A 16 ft wide, rectangular channel goes from a very steep grade to a mild slope. The design discharge is 875 ft³/s and the normal depth and velocity on the steep slope were calculated to be 1.0 ft and 49 ft/s, respectively. On the mild slope, the normal depth and velocity were calculated to be 9.71 ft and 5.51 ft/s, respectively. Determine the type of flow occurring in both channels. If a hydraulic jump occurs, evaluate the depth downstream of the hydraulic jump, the location of the jump, and the water surface profile classification.

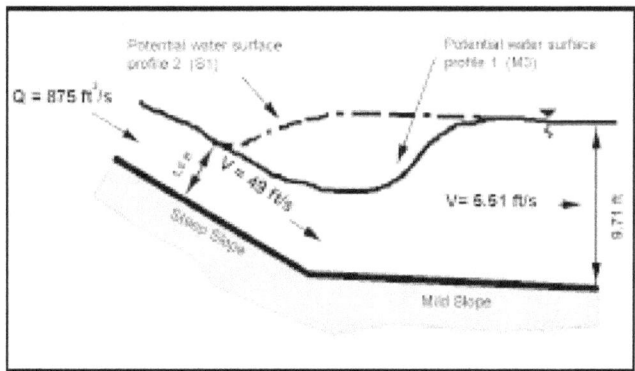

Find:

1.   Find the critical depth, y_c, on the steep slope.

$$y_c = \left(\frac{q^2}{g} \right)^{\frac{1}{3}} \qquad q = Q/B \text{ where B - channel width}$$

$$y_c = \left[\frac{(Q/B)^2}{g} \right]^{\frac{1}{3}} = \left[\frac{(875/16)^2}{32.2} \right]^{\frac{1}{3}} = 4.52 \text{ ft}$$

On the steep slope, the normal depth is 1.0 ft. Since $y < y_c$, supercritical flow occurs on the steep slope. Note that the unit discharge, q, is the same for the mild slope and hence, y_c, is the same for the steep and mild slope sections. On the mild slope, the normal depth is 9.71 ft. Since $y > y_c$, subcritical flow occurs on the mild slope. Therefore, a hydraulic jump should occur.

2.   Next, determine if the jump will occur on the steep slope or on the mild slope.

To determine if the hydraulic jump occurs on the steep or mild slope, calculate the sequent depth (y_2) for the steep slope y_1 depth using the hydraulic jump equation. If y_2 from the hydraulic jump is larger than the normal depth y_0 from Manning's equation on the mild slope, then there will be an M3 curve on the mild slope until the y_2 equals the critical depth. If y_2 is smaller than y_0 on the mild slope, then the jump may occur on the steep slope and an S1 curve will occur to connect with the normal depth at the control section.

$$y_2 = \frac{y_1}{2}\left(\sqrt{1 + 8\,Fr_1^2} - 1\right)$$

$$Fr = \frac{V}{\sqrt{gy}} = \frac{49}{\sqrt{32.2\,(1.0)}} = 8.64$$

$$y_2 = \frac{1.0}{2}\left(\sqrt{1 + 8\,(8.64)^2} - 1\right) = 11.72 \text{ ft}$$

Compare parameters and determine the type of water surface classification.

Since the sequent depth (y_2) is greater than the mild slope normal depth (i.e., 11.72 > 9.71), the mild slope channel will have an M3 curve until the hydraulic jump occurs.

4.7.3. Overview of Calculation Procedure

The standard step method is a simple computational procedure to determine the water surface profile in gradually varied flow. Prior knowledge of the type of water surface profile as determined in the preceding section would be useful to determine whether the analysis should proceed up- or downstream.

The standard step method is derived from the energy equation. The equation is

$$\Delta L = \frac{H_1 - H_2}{S_0 - S_f}$$

where:

ΔL	=	Distance between sections 1 and 2, m (ft)
H	=	Specific energy at sections 1 and 2, m (ft)
S_0	=	Slope of the bed
S_f	=	Friction slope

The above equation is used in the standard step method. An example of the use of the standard step method is given in manuals such as HIRE.[11]

Although computer programs such as WSPRO and HY8 are now used to compute water surface profiles, it is recommended that a qualitative sketch of the water surface profiles be made using the information given in the preceding section. This is particularly useful in complicated profiles where the channel slopes change from steep to mild or mild to steep.

5. APPLICATIONS OF OPEN-CHANNEL FLOW

5.1. General Design Concepts

The capacity of a drainage channel depends upon its shape, size, slope, and roughness. For a given channel, the capacity becomes greater when the grade or the depth of flow is increased. The channel capacity decreases as the channel surface becomes rougher. For example, a riprap-lined ditch has only about half the capacity of a concrete-lined ditch of the same size, shape, and slope because of the differences in channel roughness. A rough channel is sometimes an advantage on steep slopes where it is desirable to keep velocities from becoming too high.

The most efficient shape of channel is that of a semi-circle, but hydraulic efficiency is not the sole criterion. In addition to performing its hydraulic function, the drainage channel should be economical to construct and require little maintenance during the life of the roadway. Channels should also be safe for vehicles accidentally leaving the traveled way, pleasing in appearance, and dispose of collected water without damage to the adjacent property. Most of these additional requirements for drainage channels reduce the hydraulic capacity of the channel. Thus, the best design for a particular section of highway is a compromise among the various requirements, sometimes with each requirement having a different influence on the design. Figure 27 illustrates the preferred geometric cross section for ditches with gradual slope changes in which the front and back slopes are traversable.[25] This figure is applicable for rounded ditches with bottom widths of 2.4 m (8 ft) or more, and trapezoidal ditches with bottom widths equal to or greater than 1.2 m (4 ft).

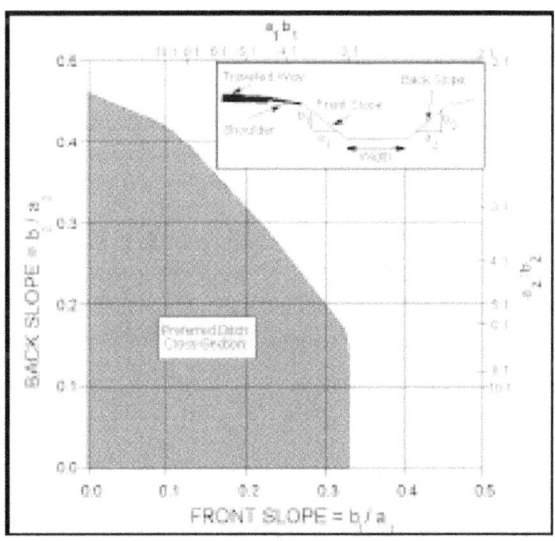

Figure 27. Preferred ditch cross section geometry.[25]

The width of the right-of-way usually allows little choice in the alignment or in the grade of the channel, but as far as practicable, abrupt changes in alignment or in grade should be avoided. A sharp change in alignment presents a point of attack for the flowing water and abrupt changes in grade cause deposition of transported material when the grade is flattened or scour when grade is steepened.

It is unnecessary to standardize the design of roadway drainage channels for any length of the highway. Not only can the depth and breadth of the channel be varied with variation in the amounts of runoff, channel grade, and distance between lateral outfall culverts, but the dimensions can be varied by the use of different types of channel lining. Nor is it necessary to standardize the lateral distance between the channel and the edge of pavement. Often liberal offsets can be obtained where cuts are slight and where cuts end and fills begin.

Systematic maintenance is essential to any drainage channel. Without proper maintenance, a well-designed channel can become an unsightly gully. Maintenance methods should be considered in the design of drainage channels so that the channel sections will be suitable for the methods and equipment that will be used for their maintenance (see chapter 10).

5.2. Stable Channel Design

5.2.1. Basic Concepts

Roadside channels are commonly used with uncurbed roadways to convey runoff from the highway pavement, and from areas that drain toward the highway. A channel section may also be used with curbed highway sections to intercept off-pavement drainage in order to minimize deposition of sediment and other debris on the roadway, and to reduce the amount of water that must be carried by the roadway section. The gradient of roadside channels typically parallels the grade of the highway. Even at relatively mild highway grades, highly erosive hydraulic conditions can exist in adjacent roadside channels. Consequently, designing a stable conveyance becomes a critical component in the design of roadside channels.

The need for erosion prevention is not limited to the highway drainage channels; it extends throughout the right-of-way and is an essential feature of adequate drainage design. Erosion and maintenance are minimized largely by the use of flat sideslopes rounded and blended with natural terrain, drainage channels designed with due regard to location, width, depth, slopes, alignment, and protective treatment, proper facilities for groundwater interception, dikes, berms, and other protective devices, and protective ground covers and planting.

The discussion in this chapter is limited to providing erosion control in drainage channels by proper design, including the selection of an economical channel lining. Lining as applied to drainage channels includes vegetative coverings. The type of lining should be consistent with the degree of protection required, overall cost, safety requirements, and esthetic considerations. Control of erosion caused by overland or sheet flow is not discussed.

5.2.2. Lining Materials

Lining materials may be classified as flexible or rigid. Flexible linings are able to conform to changes in channel shape and can sustain such changes while maintaining the overall integrity of the channel. In contrast, rigid linings cannot change shape and tend to fail when a portion of the channel lining is damaged. Channel shape may change due to frost-heave, slumping, piping, etc. Typical flexible lining materials include grass and riprap (figures 28 and 29), while a typical rigid lining material is concrete (figure 30). Flexible linings are generally less expensive, have a more natural appearance, and are typically more environmentally acceptable. However, flexible linings are limited in the erosive forces they can sustain without damage to the channel and lining. A rigid lining can typically provide higher capacity and greater erosion resistance, and in some cases may be the only feasible alternative.

96

Figure 28. Vegetative channel lining.

Figure 29. Riprap channel lining.

Figure 30. Rigid concrete channel lining.

Flexible linings can be either permanent or temporary. Temporary flexible linings are commonly used to provide erosion protection until a permanent lining, such as grass, is established. Temporary flexible linings are typically a net or mat of either natural or synthetic fibers that is laid in the channel after seeding operations and secured with staples or stakes (figure 31). Permanent flexible linings include vegetation and rock riprap. Vegetative linings may consist of planted or sodded grasses placed in and along the drainage. Rock riprap is dumped in place on a filter (granular or geotextile), on a prepared slope to form a well-graded mass with a minimum of voids. Rocks should be hard, durable, preferably angular in shape and free from overburden, shale and organic material. Resistance to disintegration from channel erosion should be determined from service records or from specified field or laboratory tests.

Figure 31. Installed synthetic mat channel lining.

Construction of rigid concrete linings requires specialized equipment and costly materials. As a result the cost of rigid linings is typically high. Prefabricated linings can be a less expensive alternative if shipping distances are not excessive. Interlocking concrete paving blocks are a typical prefabricated lining.

In general, when a lining is needed, the lowest cost lining that affords satisfactory protection should be used. In humid regions, this is often vegetation used alone or in combination with other types of linings. Thus, a channel might be grass-lined on the flatter slopes and lined with more resistant material on the steeper slopes. In cross section, the channel might be lined with a highly resistant material within the depth required to carry floods occurring frequently and lined with grass above that depth for protection from the rare floods.

5.2.3. Stable Channel Design

Stable channel design can be based on the concepts of static or dynamic equilibrium. Static equilibrium exists when the channel boundaries are essentially rigid and the material forming the channel boundary effectively resists the erosive forces of the flow. Under such conditions the channel remains essentially unchanged during the design flow and the principles of rigid boundary hydraulics can be applied. Dynamic equilibrium exists when the channel boundary is moveable and some change in the channel bed and/or banks occurs. A dynamic system is considered stable as long as the net change does not exceed acceptable levels. Designing a stable channel under dynamic equilibrium conditions must be based on the concepts of sediment transport. For most highway drainage channels bed and bank instability and/or possible lateral migration cannot be tolerated and stable channel design must be based on the concepts of static equilibrium, including the use of a lining material if necessary to achieve a rigid boundary condition.

Two methods have been developed and are commonly applied to design static equilibrium channel conditions: the permissible velocity approach and the permissible tractive force (shear stress) approach. Under the permissible velocity approach the channel is assumed stable if the adopted mean velocity is lower than the maximum permissible velocity for the given channel boundary condition. Similarly, the tractive force approach requires that the shear stresses on the channel bed and banks do not exceed the allowable amounts for the given channel boundary. Permissible velocity procedures were first introduced around the 1920s and have been developed and widely used by the Soil Conservation Service. Tractive force procedures based on shear stress concepts (see chapter 4) originated largely through research by the Bureau of Reclamation in the 1950s. Based on the actual physical processes involved in maintaining a stable channel, specifically the stresses developed at the interface between flowing water and materials forming the channel boundary, the tractive force procedure is a more realistic model and was adopted as the preferred design procedure for flexible linings in Hydraulic Engineering Circular Number 15, entitled "Design of Roadside Channels with Flexible Linings" (HEC-15), which is the primary reference for stable channel design.[26]

The definition and equation for computing the tractive force was provided in chapter 4 (equation 25). This defines the average tractive force on the channel. The maximum shear stress along the channel bottom may be estimated by substituting the flow depth, y, for the hydraulic radius, R, in equation 25. The permissible tractive force for a variety of lining materials is provided by table 13. If the computed tractive force is greater than the permissible tractive force for a given lining material, the channel will not be stable. Calculation of hydraulic geometry conditions can be based on normal flow depth conditions. Table 14 provides Manning's n values for nonvegetative flexible linings. The n values will vary with depth, with greater roughness for shallow depth and less roughness at large flow depths. HEC-15 details the tractive force stable channel design procedure, including special considerations for steep-slope riprap design and design of composite linings.[26]

5.3. Pavement Drainage Design

5.3.1. Basic Concepts

Pavement drainage design provides for effective removal of water from the roadway surface. Water flowing on a pavement with a curb is essentially a special case of open-channel flow in a shallow, triangular-shaped cross section. Pavement drainage to a gutter or swale adjacent to the roadway surface is another case of open-channel flow, again typically occurring in a wide shallow cross section.

Water on the pavement will slow traffic and contribute to accidents from hydroplaning and loss of visibility from splash and spray. Pavement drainage design is typically based on a design discharge and an allowable spread of water across the pavement. Spread on traffic lanes can be tolerated more frequently and to greater widths where traffic volumes and speeds are low. In contrast, high speed, high volume facilities, such as freeways, should be designed to minimize or eliminate spread of water on the traffic lanes during the design event.

EXAMPLE PROBLEM 5.1 (SI Units)

Determine whether it is feasible to use jute net as a temporary lining.

Given: Q = 0.6 m³/s
S = 0.005 m/m
Trapezoidal channel with a bottom width of 1.0 m and 1V:3H sideslopes.

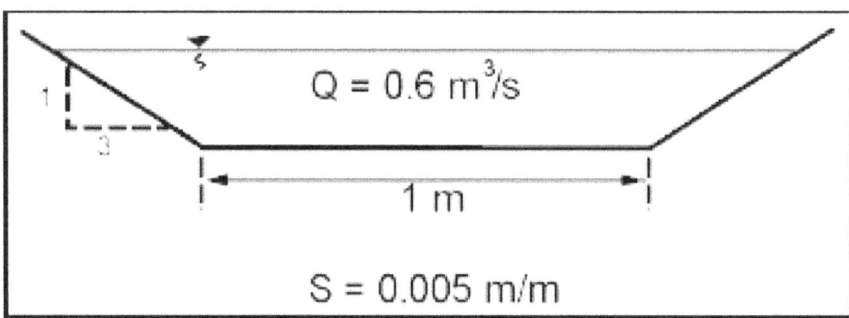

Find:

Depth of flow in the channel and the adequacy of the jute net lining.

Solution:

(1)   From table 13, the permissible shear stress is 21.6 Pa and from table 14, the Manning's n value is 0.022 (assuming a flow depth between 15-60 cm).

(2)   Solving Manning's equation for S = 0.005, Q = 0.6 m³/s, and B = 1.0, the flow depth y_s is 0.3 m

Note that the flow depth is within the assumed range of 15 to 60 cm so that the selected Manning's n value is correct, according to table 14.

(3)   The maximum shear stress on the channel bed is (equation 25 using y instead of R):

$$\tau_o = \gamma y S = 9800 \text{ N/m}^3 \times 0.3 \times 0.005 = 14.7 \text{ N/m}^2 = 14.7 \text{ Pa}$$

(4)   Comparing the shear stress, 14.7 Pa, to the permissible shear stress, 21.6 Pa, shows that jute net is an acceptable channel lining.

EXAMPLE PROBLEM 5.1 (English Units)

Determine whether it is feasible to use jute net as a temporary lining.

Given: $Q = 20$ ft^3/s
    $S = 0.005$ ft/ft
    Trapezoidal channel with a bottom width of 3 ft and 3:1 sideslopes.

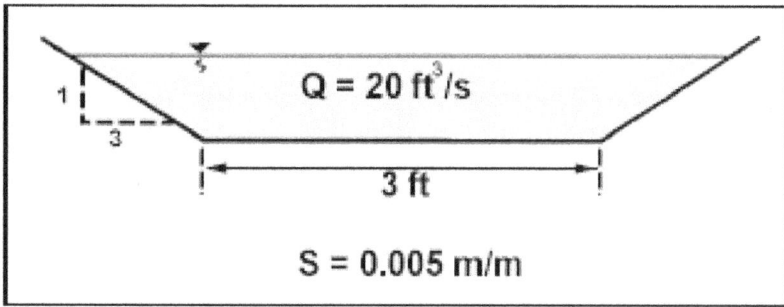

$Q = 20$ ft^3/s

3 ft

$S = 0.005$ m/m

Find:

Depth of flow in the channel and the adequacy of the jute net lining.

Solution:

(1)   From table 13, the permissible shear stress is 0.45 lb/ft^2 and from table 14, the Manning's n value is 0.022 (assuming a flow depth between 0.5-2.0 ft).

(2)   Solving Manning's equation for $S = 0.005$, $Q = 20$ ft^3/s, and $B = 3$ ft, the flow depth y_s is 0.97 ft

    Note that the flow depth is within the assumed range of 0.5 to 2.0 ft so that the selected Manning's n value is correct, according to table 14.

(3)   The maximum shear stress on the channel bed is (equation 25 using y instead of R):

$$\tau_o = \gamma y S = 62.4 \, lb/ft^3 \; 0.97 \, ft \times 0.005 = 0.30 \, lb/ft^2$$

(4)   Comparing the shear stress, 0.30 lb/ft^2, to the permissible shear stress, 0.45 lb/ft^2, shows that jute net is an acceptable channel lining.

Standard roadway geometric design features greatly influence pavement drainage design. These features include curbs, gutter configuration, longitudinal and lateral pavement slope, shoulders and parking lanes. Curbing at the right edge of pavements is normal practice for low-speed, urban highway facilities. Gutters adjacent to the curb combined with a portion of the shoulder or roadway pavement, depending on allowable spread, are used to carry runoff.

Longitudinal grade influences the spread of water onto the pavement. Curbed pavements typically require a minimum slope of 0.3 percent to promote drainage. Minimum grades can be maintained in very flat terrain by use of rolling profiles or by warping the cross slope to achieve a rolling gutter profile. Pavement cross slope is often a compromise between the need for reasonably steep cross slopes for drainage and relatively flat cross slopes for driver comfort. Adequate cross slope will reduce water depth on the pavement and, therefore, is an important countermeasure against hydroplaning.

In areas where vegetative cover cannot be used to prevent erosion damage to high fills, shoulders are designed to serve as a gutter with a curb constructed at the outer edge to confine the water to the shoulder. The water collected in the gutter is discharged down the slope through down drains. The curb may be made of bituminous or portland cement concrete.

The primary design reference for highway pavement drainage design is HEC-22.[6] The following information has been extracted from that document and provides an overview of pavement drainage design concepts and considerations. For more detailed information and design criteria, see reference 6.

5.3.2. Flow in Gutters and Swales

The triangular shaped area defined by the curb, gutter and the spread onto the pavement creates an open-channel flow section for conveying runoff (figure 32). Gutters adjacent to the curb can have a steeper cross slope from the pavement. This steeper gutter section can be an effective countermeasure for reducing spread on the pavement.

Modification of Manning's equation (equation 13) is necessary for use in computing flow in triangular channels because the hydraulic radius in the equation does not adequately describe the gutter cross section, particularly where the top width of the water surface may be more than 40 times the depth at the curb. To compute gutter flow Manning's equation is integrated for an increment of width across the section, and the resulting equation in terms of cross slope and spread is (HEC-22):[6]

$$Q = \frac{K_u}{n} S_x^{5/3} S^{1/2} T^{8/3} \tag{50}$$

where:

Q	=	Discharge in m³/s (ft³/s)
S_x	=	Cross slope in m/m (ft/ft)
S	=	Longitudinal slope in m/m (ft/ft)
T	=	Spread in m (ft)
K_u	=	0.376 (0.56)

Note that nomographs for solution of equation 50, for both uniform and for the more complex geometry created by composite cross slopes, are provided in HEC-22.[6] Tabulated flow capacity for standard highway cross sections may also be available from local and regional design guides.

Swale sections at the edge of the roadway pavement or shoulder offer advantages over curbed sections where curbs are not needed for traffic control. These advantages include less hazard to traffic than a near-vertical curb and hydraulic capacity that is not dependent on spread on the pavement. Swale sections are particularly appropriate where curbs are used to prevent water from eroding fill slopes. Swale sections are typically circular or v-shape. HEC-22 also provides nomograph solution of Manning's equation for shallow swale sections.[6]

102

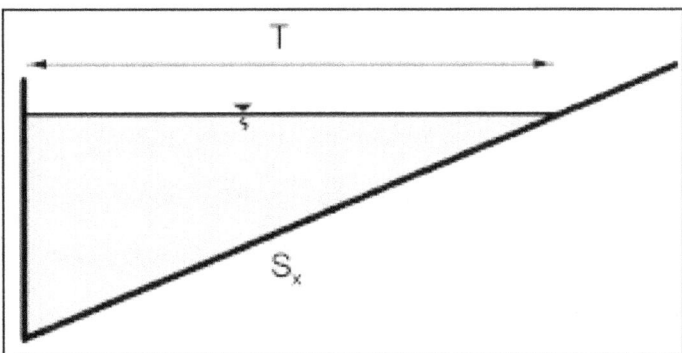

Figure 32. Definition sketch - triangular section.

EXAMPLE PROBLEM 5.2 (SI Units)	EXAMPLE PROBLEM 5.2 (English Units)
Given: A gutter with the following dimensions and conditions	Given: A gutter with the following dimensions and conditions
$T = 2.5$ m $\\ S_x = 0.02 \\ S = 0.01 \\ n = .016$	$T = 8.0$ ft $\\ S_x = 0.02 \\ S = 0.01 \\ n = 0.016$
Find: Flow in gutter at design spread	Find: Flow in gutter at design spread
Solution: (use the modified Manning's equation)	Solution: (use the modified Manning's equation)
$$Q = \frac{K_u}{n} S_x^{5/3} S^{1/2} T^{8/3}$$	$$Q = \frac{K_u}{n} S_x^{5/3} S^{1/2} T^{8/3}$$
$$Q = \frac{.376}{0.015}(0.02)^{5/3}(0.01)^{1/2}(2.5)^{8/3}$$	$$Q = \frac{0.56}{0.016}(0.02)^{5/3}(0.01)^{1/2}(8.0)^{8/3}$$
$Q = 0.040$ m^3/s	$Q = 1.31$ ft^3/s

5.3.3. Pavement Drainage Inlets

When the capacity of the curb/gutter/pavement section has been exceeded, typically as a result of spread considerations, runoff must be diverted from the roadway surface. A common solution is often interception of all or a portion of runoff by drainage inlets that are connected to a storm drain pipe. Inlets used for intercepting runoff from highway surfaces can be divided into four major classes: (1) curb-opening inlets, (2) grate inlets, (3) slotted drains, and (4) combination inlets (figures 33 and 34). Each class has many variations in design and may be installed with or without a depression of the gutter.

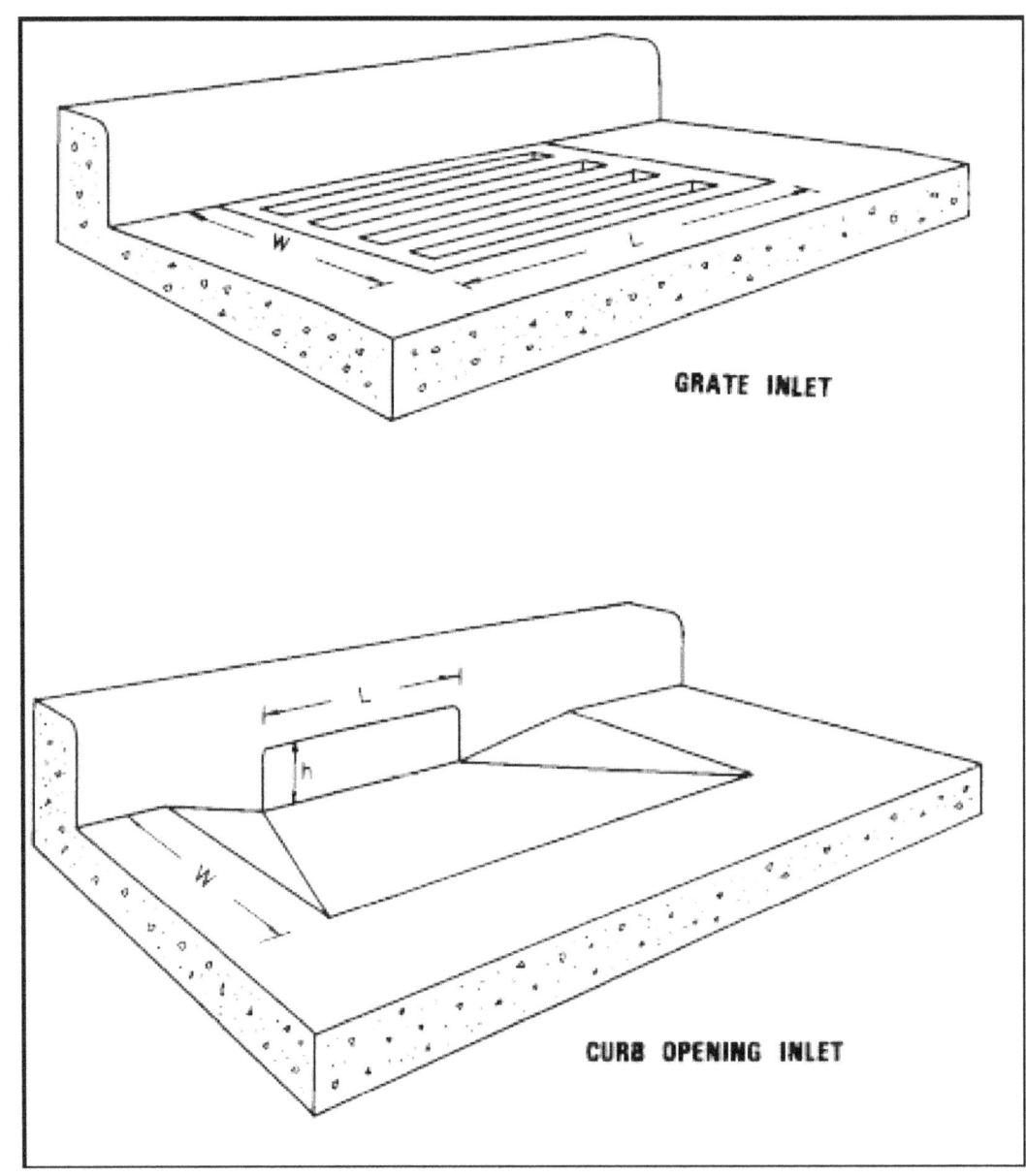

Figure 33. Perspective views of grade and curb-opening inlets.

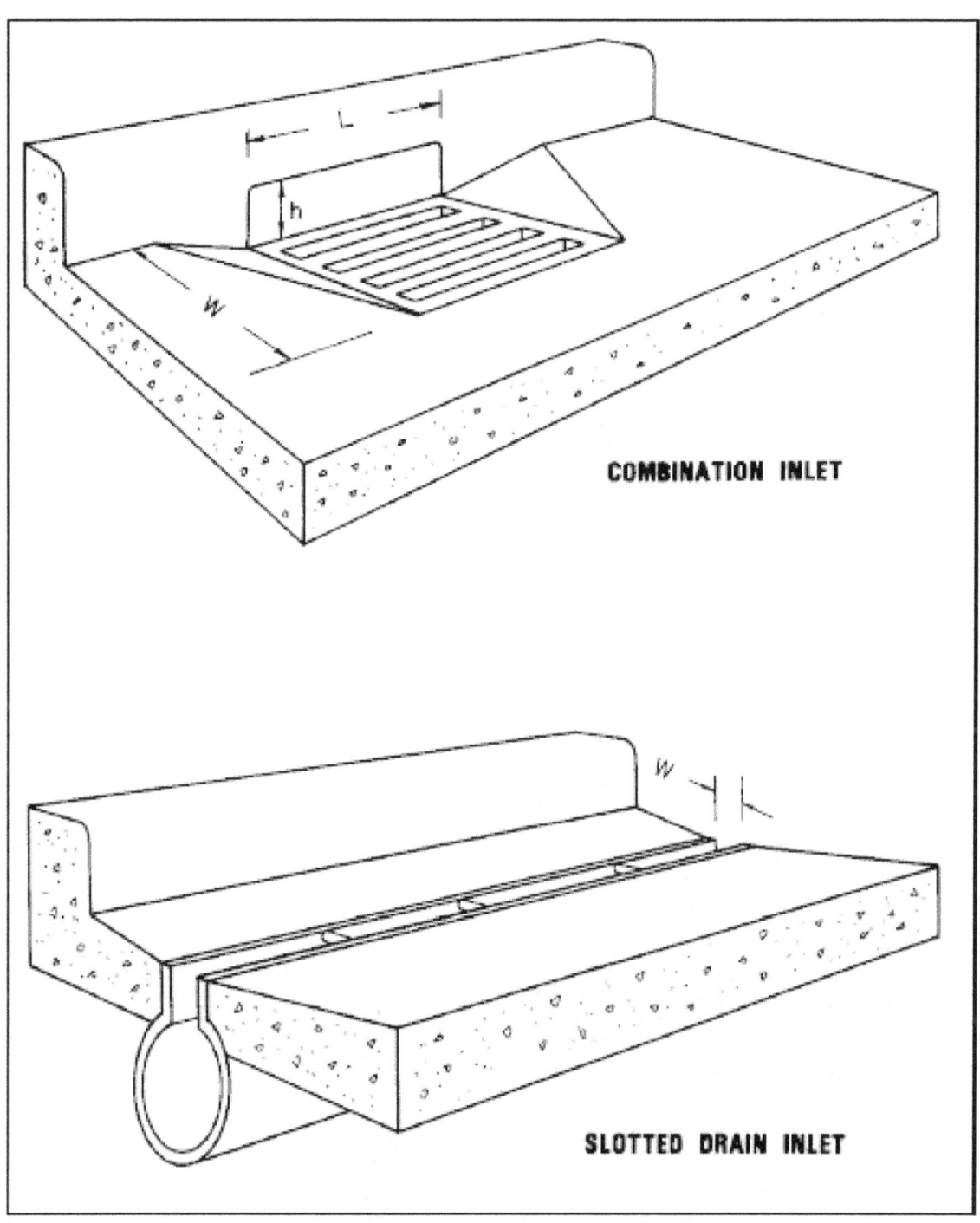

Figure 34. Perspective views of combination inlet and slotted drain inlet.

Inlet capacity is a function of a variety of factors, including type of inlet, grate design, location (on grade or in a sag location), gutter design, debris clogging, etc. Inlets on continuous grade operate as weir flow, while inlets in sag locations will initially operate as weir flow, but will transition to orifice flow as depth increases (see discussion of weirs and orifices in chapter 3). Orifice flow begins at depths dependent on the grate size, the curb opening height, or the slot width of the inlet, depending on the type of inlet/grate. At depths between those where weir flow occurs and those where orifice flow occurs, flow is in a transition stage and may be ill-defined and poorly behaved.

The efficiency of inlets in passing debris is critical in sag locations since all runoff that enters the sag must pass through the inlet, otherwise hazardous ponding condition can result. Grate inlets alone are not recommended in sag locations because of potential clogging.

Inlet capacity is typically defined by design charts developed for standard inlet configurations from laboratory testing. As an example, the design chart for total interception by curb-inlets and slotted drains is shown in figure 35. Figure 36 is the interception efficiency for the same inlets when less than total interception occurs. Example problem 5.3 illustrates the use of these design charts. HEC-22 provides other design charts for a wide range of inlet and grate types typically used in highway engineering.[6]

5.3.4. Median, Embankment, and Bridge Inlets

Medians are commonly used to separate opposing lanes of traffic on divided highways. Median areas should preferably not drain across traveled lanes, and often times the inside lanes and shoulder of multi-lane highways will drain to the median area where a center swale collects the runoff. Based on capacity or erosion considerations, it is sometimes necessary to place inlets in medians to remove some or all the runoff that has been collected. Medians may be drained by drop (grate) inlets similar to those used for pavement drainage (figure 37).

Effective bridge deck drainage is important for several reasons, including hydroplaning, ice formation, and susceptibility of the deck structural and reinforcing steel to corrosion from deicing salts. While bridge deck drainage is accomplished in the same manner as any other curbed roadway section (figure 38), bridge decks are often less effectively drained because of lower cross slopes, uniform cross slopes for traffic lanes and shoulders, parapets that collect debris, and drainage inlets that are relatively small and susceptible to clogging. Because of the limitations of bridge deck drainage, roadway drainage should be intercepted where practical before it reaches a bridge.

Drainage inlets used to intercept runoff upgrade or downgrade of bridges, or runoff that might endanger an embankment fill slope, differ from other pavement drainage inlets. First, the economies achieved by system design are not possible because a series of inlets are not used. Second, total or near total interception is necessary and third, a closed storm drain system is often not available to dispose of the intercepted flow. Intercepted flow is usually discharged into open chutes or pipe downdrains terminating at the toe of the fill slope (figure 39).

5.4. Open-Channel Flow Analysis Using HYDRAIN

Evaluation of the gradually varied flow water surface profile in an open channel can be completed with the Water Surface Profile (WSPRO) Program in HYDRAIN.[10] WSPRO was designed to provide a water surface profile for six major types of open-channel flow situations: (1) unconstricted flow, (2) single opening bridge, (3) bridge opening(s) with spur dikes, (4) single opening embankment overflow, (5) multiple alternatives for a single job, and (6) multiple openings. WSPRO is a powerful analytical tool, particularly for the analysis and design of bridge openings. The HYDRA module of HYDRAIN will analyze pavement drains for grates, curb openings, combinations, and slotted drains. It will also design or analyze the storm drain system which operates as open-channel flow in many instances.

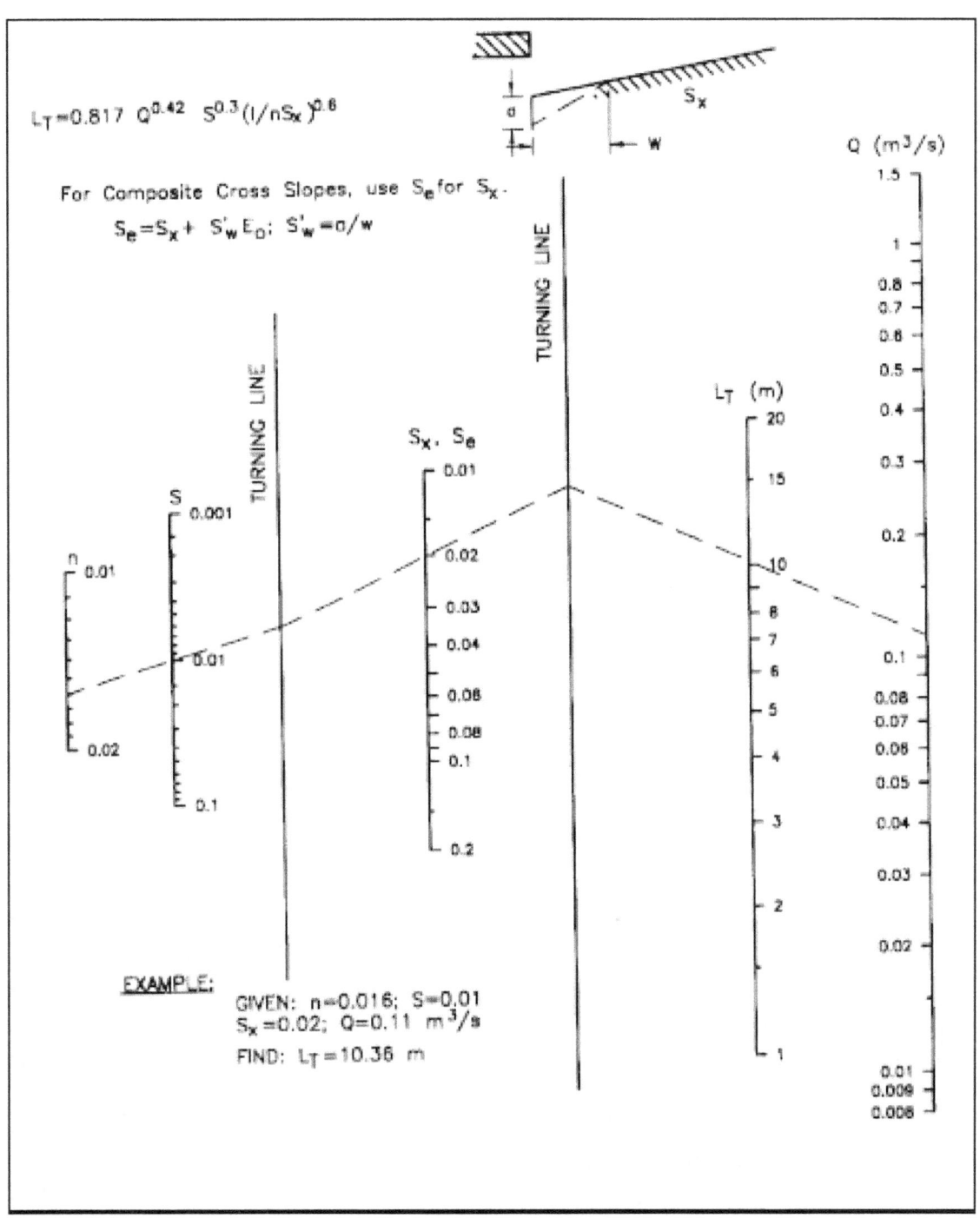

Figure 35a. SI chart for curb-opening and slotted-drain inlet length for total interception.

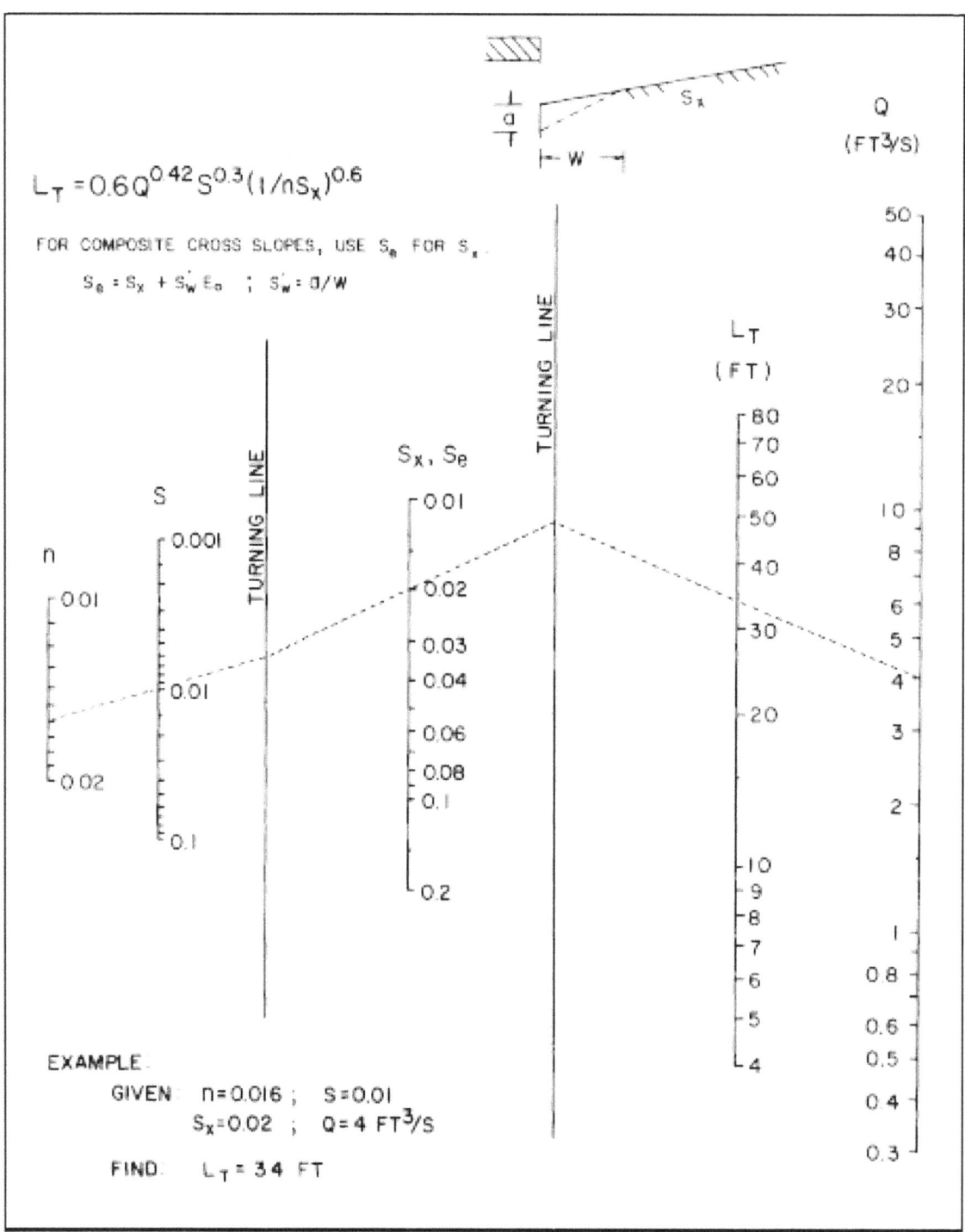

Figure 35b. English chart for curb-opening and slotted-drain inlet length for total interception.

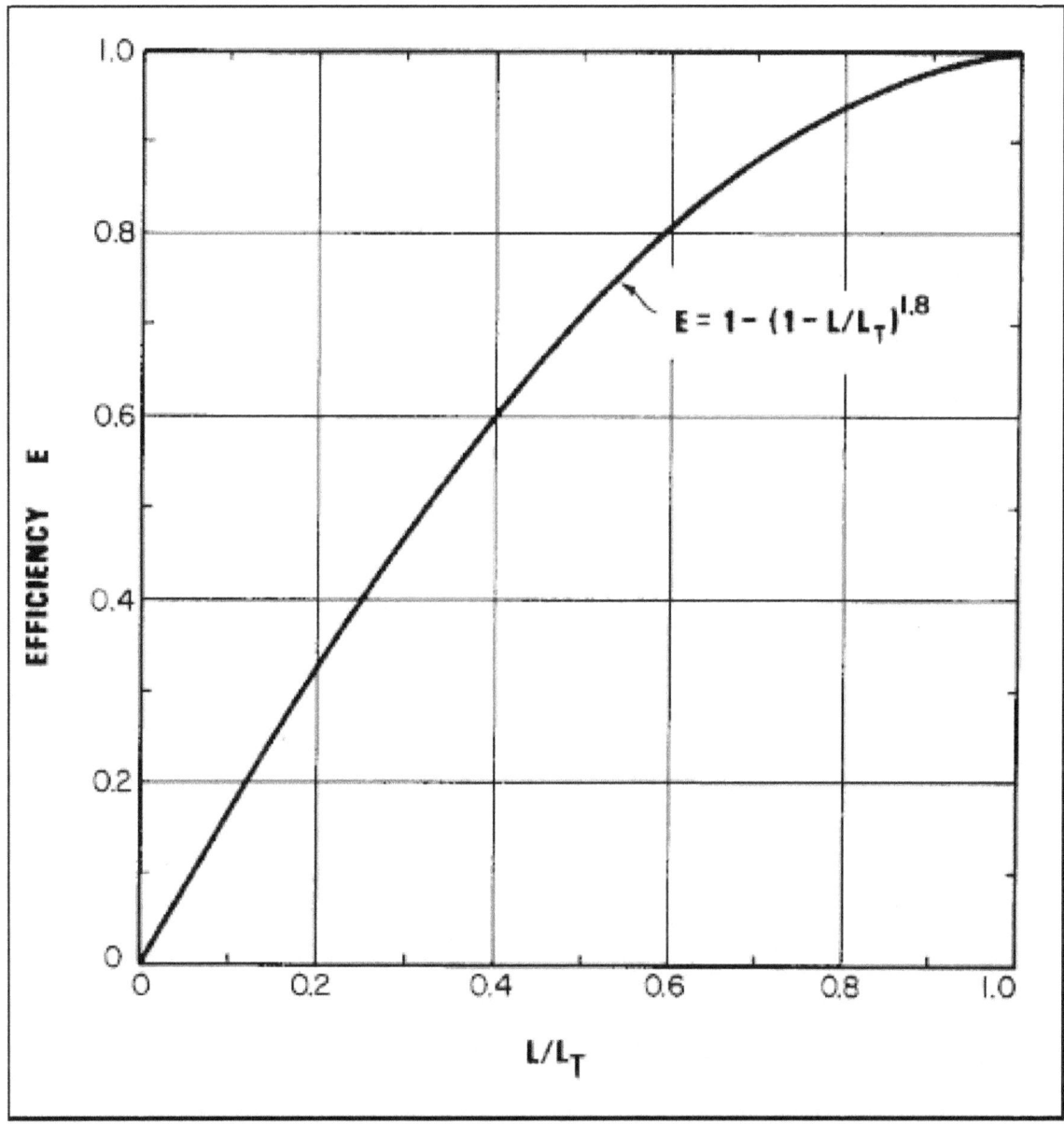

Figure 36. Curb-opening and slotted-drain inlet interception efficiency.[6]

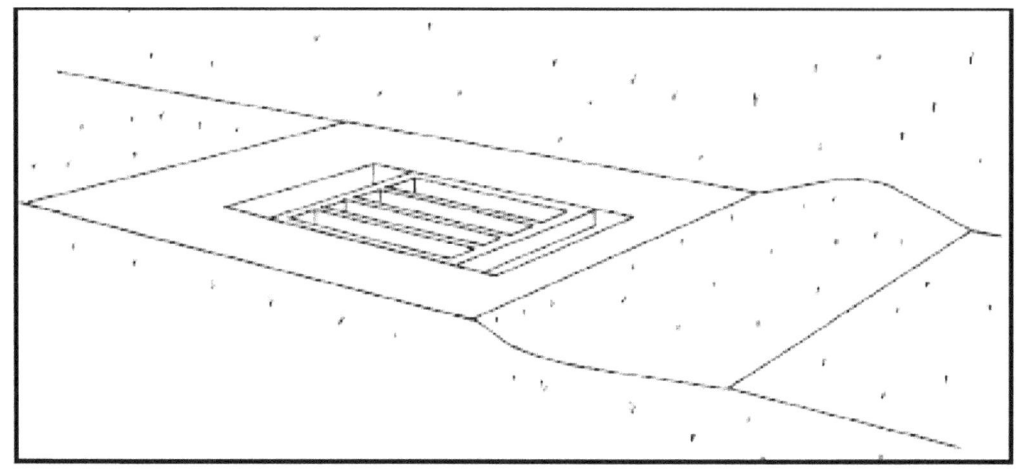

Figure 37. Median drop inlet.

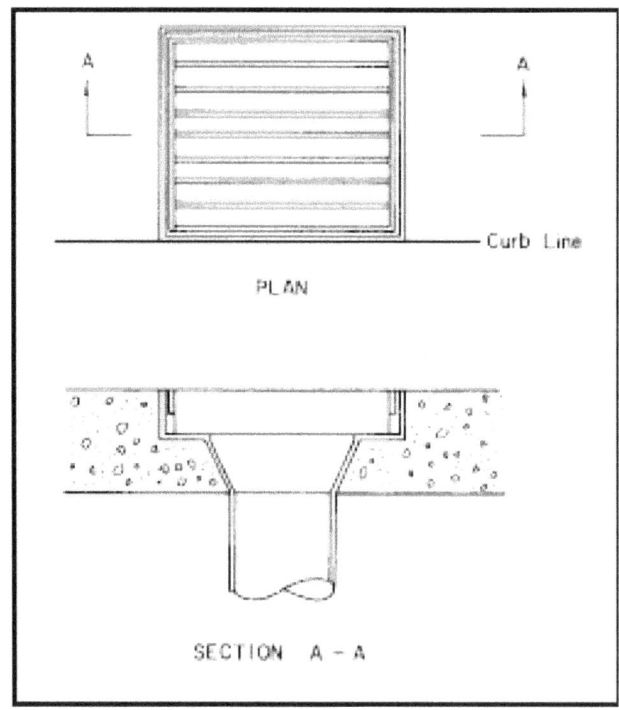

Figure 38. Bridge inlet.

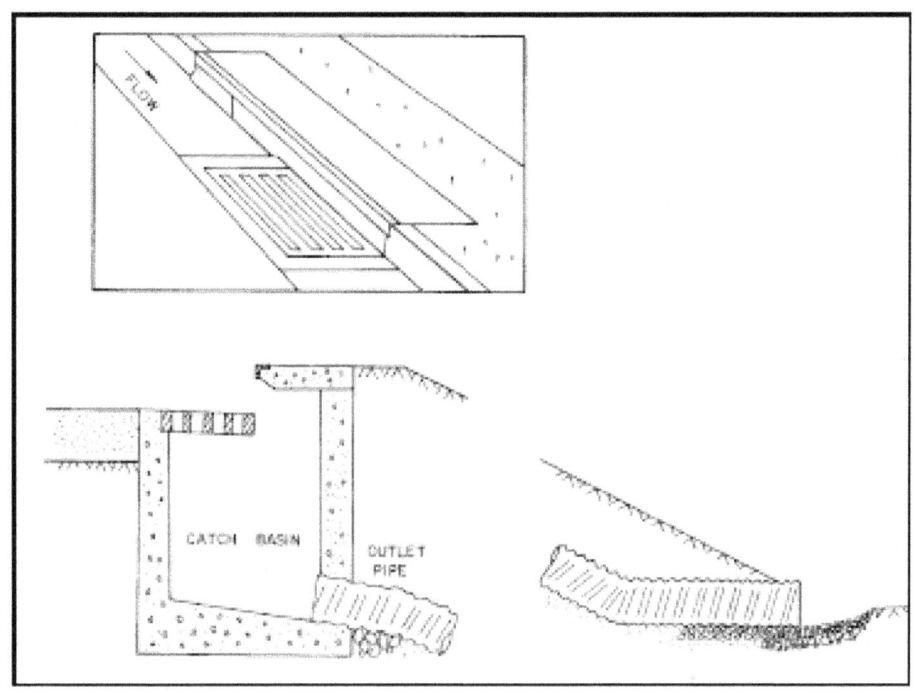

Figure 39. Embankment inlet and downdrain.

EXAMPLE PROBLEM 5.3 (SI Units)

Design the length of curb opening inlet required to intercept 0.07 m³/s flowing along a street with a cross slope of 0.02, and a longitudinal slope of 0.03. Assume an n value of 0.016. Also determine the discharge intercepted if a 3.0 m curb inlet is used, and the amount of bypass flow to the next inlet.

Given:

$$S_x \quad = \quad 0.2$$
$$S \quad = \quad 0.030$$
$$Q_I \quad = \quad 0.07 \text{ m}^3/\text{s}$$
$$n \quad = \quad 0.016$$

Find:

(1)   Required length for total interception by a curb-opening inlet.

(2)   Discharge intercepted by a 3.0 m curb-opening inlet, and the amount of bypass flow at the next inlet.

Solution:

(1)   From figure 35, for the given conditions a curb opening inlet 11.7 m long would intercept the total design flow of 0.07 m³/s.

(2)   If a 3.0 m curb-opening inlet is used, only a portion of the design flow will be intercepted. Figure 36 defines the interception efficiency for curb-opening inlets based on the length for total interception (L_t).

Therefore, given a 11.7 m length for total interception from item (1), and an a curb-opening length of only 3.0 m, the ratio of L/L_t is

L/L_t= 3.0/11.7 = 0.26

From figure 36 the efficiency, E, is then 0.41, and the discharge intercepted by a 3.0 m curb-opening inlet is

$Q_i = EQ = (0.41) (0.07 \text{m}^3/\text{s}) = 0.029 \text{ m}^3/\text{s}$

The amount of bypass flow to the next inlet is

0.07 - 0.029 - 0.041 m³/s

112

EXAMPLE PROBLEM 5.3 (English Units)

Design the length of curb opening inlet required to intercept 2.29 ft^3/s flowing along a street with a cross slope of 0.02, and a longitudinal slope of 0.03. Assume an n value of 0.016. Also determine the discharge intercepted if a 10 ft curb inlet is used, and the amount of bypass flow to the next inlet.

Given:

$$
\begin{aligned}
Sx &= .02 \\
S &= .03 \\
Q &= 2.29 \text{ ft}^3/\text{s} \\
n &= 0.016
\end{aligned}
$$

Find:

(1)   Required length for total interception by a curb-opening inlet.

(2)   Discharge intercepted by a 10-foot long curb-opening inlet, and the amount of bypass flow at the next inlet.

Solution:

(1)   From figure 35, for the given conditions a curb opening inlet 37.1 ft long would intercept the total design flow of 2.29 ft^3/s.

(2)   If a 10.0 ft curb-opening inlet is used, only a portion of the design flow will be intercepted. Figure 36 defines the interception efficiency for curb-opening inlets based on the length for total interception (L$_t$).

Therefore, given a 37.1 ft length for total interception from item (1), and an a curb-opening length of only 10 ft, the ratio of L/L$_t$ is

$$ L/L_t = \frac{10}{37.1} = 0.27 $$

From figure 36 the efficiency, E, is then 0.43, and the discharge intercepted by a 10 ft curb-opening inlet is

Qi = EQ = (0.43) (2.29ft^3/s) = 0.99

The amount of bypass flow to the next inlet is

2.29 - .99 - 1.30 ft^3/s

(page intentionally left blank)

6.   CLOSED-CONDUIT FLOW

6.1.   Types of Flow in Closed Conduits

Flow conditions in a closed conduit can occur as open-channel flow, gravity full flow or pressure flow. In open-channel flow the water surface is exposed to the atmosphere, which can occur in either an open conduit or a partially full closed conduit. The analysis of open-channel flow in a closed conduit is no different than any other type of open-channel flow, and all the concepts and principles discussed in chapter 4 are applicable. Gravity full flow occurs at that condition where the conduit is flowing full, but not yet under any pressure. Pressure flow occurs when the conduit is flowing full and under pressure.

Due to the additional wetted perimeter and increased friction that occurs in a gravity full pipe, a partially full pipe will actually carry greater flow. For a circular conduit the peak flow occurs at 93 percent of the height of the pipe, and the average velocity flowing one-half full is the same as gravity full flow (figure 40). Gravity full flow condition is usually assumed for purposes of storm drain design. The Manning's equation (equation 13) combined with the continuity equation (equation 3) for circular section flowing full can be rewritten as:

$$Q = \frac{K_u}{n} D^{\frac{8}{3}} S^{\frac{1}{2}} \tag{51}$$

where:

Q	=	Discharge, m^3/s (ft^3/s)
n	=	Manning's coefficient
D	=	Pipe diameter, m (ft)
S	=	Slope, m/m (ft/ft)
K_u	=	0.312 (0.46)

This equation allows for a direct computation of the required pipe diameter. Note that the computed diameter must be increased in size to a larger nominal dimension in order to carry the design discharge without creating pressure flow. The standard SI nominal sizes based on current English unit nominal sizes are given in table 2.

6.2.   Energy Equation

The energy equation was reviewed in chapter 3 (equation 4). In very simple terms the equation states that the energy head at any cross section must equal that in any other downstream section plus the intervening losses. The energy head is divided into three components: the velocity head, the pressure head and the elevation head. The energy grade line (EGL) represents the total energy at any given cross section. The energy losses are classified as friction losses and form losses (see section 6.3).

The hydraulic grade line (HGL) is below the EGL by the amount of the velocity head. In open-channel flow the HGL is equal to the water surface elevation in the channel, while in pressure flow the HGL represents the elevation water would rise to in a stand pipe connected to the conduit. For example, in a storm drain designed for pressure flow the HGL should be lower than the roadway elevation or water in the storm drain will rise up through inlets and access hole covers and flood the roadway. Similarly, if an open-channel flow condition in a storm drain is supercritical, care must be taken to insure that a hydraulic jump does not occur which might create pressure flow and a HGL above the roadway elevation.

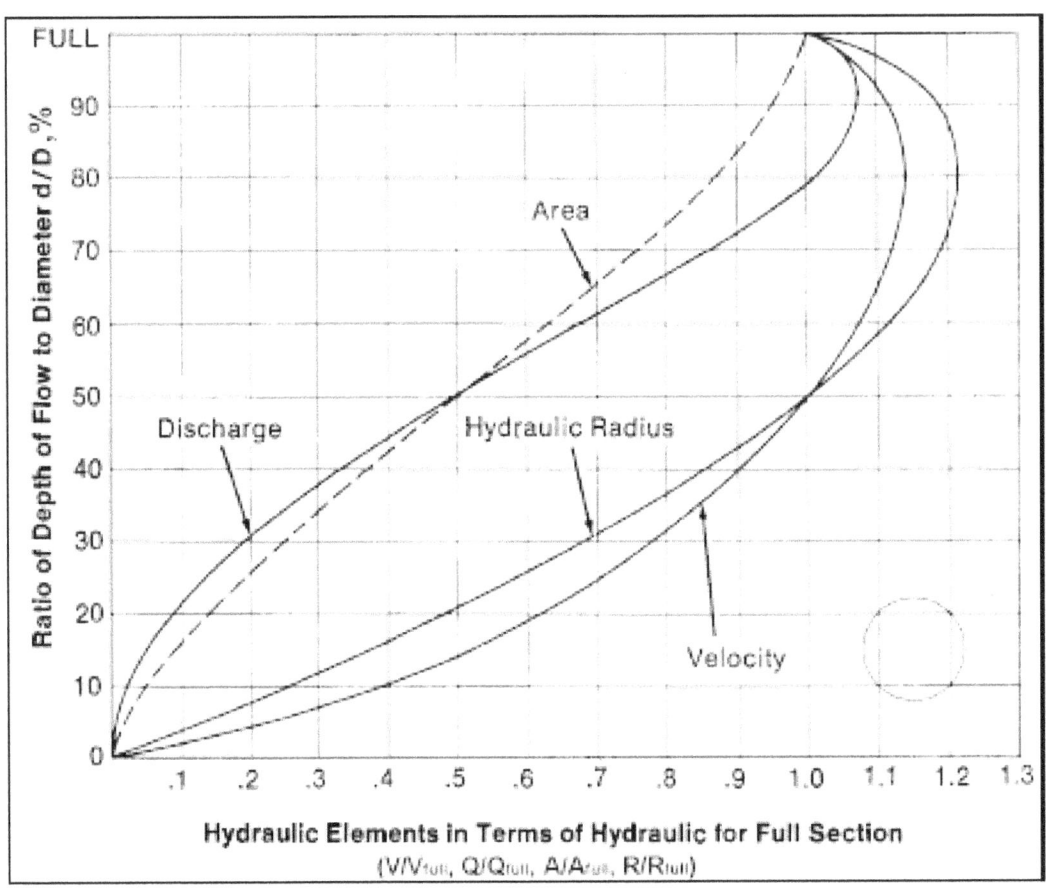

Figure 40. Part-full flow relationships for circular pipes.

6.3. Energy Losses

When using the energy equation all energy losses should be accounted for. Energy losses can be classified as friction losses or form losses. Friction losses are due to forces between the fluid and boundary material, while form losses are the result of various hydraulic structures along the closed conduit. These structures, such as access holes, bends, contractions, enlargements and transitions, will each cause velocity headlosses and potentially major changes in the energy grade line and hydraulic grade line across the structure. The form losses are often called "minor losses," which is misleading since these losses can be large relative to friction losses.

Table 2. Nominal Pipe Sizes.		
Nominal Size as Manufactured in English Units		Nominal Size Converted to SI Metric Units
Pipe Diameter		Pipe Diameter
Inches	Feet	(mm)
18	1.5	450
24	2.0	600
30	2.5	750
36	3.0	900
42	3.5	1,050
48	4.0	1,200
54	4.5	1,350
60	5.0	1,500
66	5.5	1,650
72	6.0	1,800
78	6.5	1,950
84	7.0	2,100
90	7.5	2,250
96	8.0	2,400
102	8.5	2,550
108	9.0	2,700
114	9.5	2,850
120	10.0	3,000
126	10.5	3,150
132	11.0	3,300
138	11.5	3,450
144	12.0	3,600

EXAMPLE PROBLEM 6.1 (SI Units)

Given: Pavement runoff is collected by a series of combination inlets. During the design event, the total discharge intercepted by all inlets is 0.4 m³/s. A concrete storm drain pipe (n = 0.013) is to be placed on a grade parallel to the roadway grade, which is 0.005 m/m.

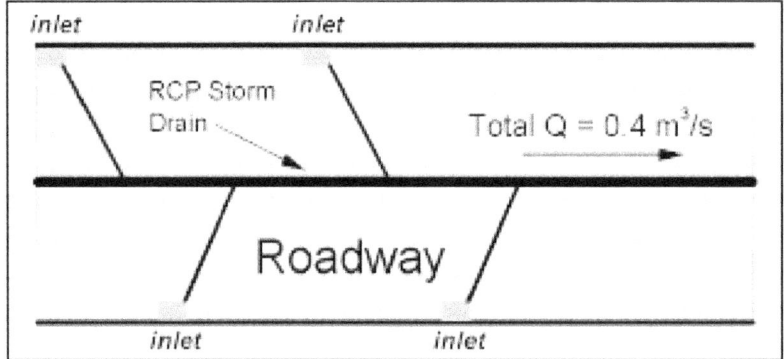

Find: The required storm drain pipe diameter and the full flow velocity.

1. Use the full flow equation (which gives pipe diameter in m)

$$Q = \frac{K_u}{n} D^{8/3} S^{1/2} \quad \text{where } K_u = 0.312 \text{ for SI units}$$

$$0.4 = \frac{0.312}{0.013} D^{8/3} (0.005)^{1/2}$$

$$D^{8/3} = 0.236$$

$$D = 0.58 \text{ m or } 580 \text{ mm}$$

2. Based on table 2, the next larger nominal pipe size is 600 mm.

3. Use the continuity equation (Q = VA) to calculate the full flow velocity.

$$Q = VA; \text{ therefore, } V = Q/A$$

$$A = \frac{\pi D^2}{4}$$

$$= \frac{\pi (0.6)^2}{4} = 0.28 \text{ m}^2$$

$$V = \frac{0.4}{0.28} = 1.4 \text{ m/s}$$

4. Note: This is the full-flow velocity; however, under our design conditions, a 600 mm would be flowing slightly less than full. Based on the part-full flow relationships (figure 40) the velocity does not change significantly from half-full to full. However, for tc calculation, the part-full velocity should be used.

To calculate the part-full velocity, nomographs or trial-and-error solution can be used. Alternatively, the part-full flow relationships can be used. The full-flow discharge and velocity of a 600-mm concrete pipe is

$$Q = \frac{0.312}{0.013}(0.60)^{8/3}(0.005)^{1/2} = 0.43 \text{ m}^3 \text{ /s}$$

$$V = \frac{Q}{A} = \frac{0.43}{(\pi(0.6)^2)/4} = 1.52 \text{ m/s}$$

The ratio of part-full to full-flow discharge is

$$\frac{Q}{Q_f} = \frac{0.40}{0.43} = 0.93$$

and from the part-full flow relationships (Figure 40), the corresponding velocity ratio is 1.13. Therefore

$$\frac{V}{V_f} = 1.13 \qquad \text{and } V = 1.13(1.52) = 1.72 \text{ m/s}$$

EXAMPLE PROBLEM 6.1 (English Units)

Given: Pavement runoff is collected by a series of combination inlets. During the design event, the total discharge intercepted by all inlets is 14.13 ft³/s. A concrete storm drain pipe (n = 0.013) is to be placed on a grade parallel to the roadway grade, which is 0.005 ft/ft.

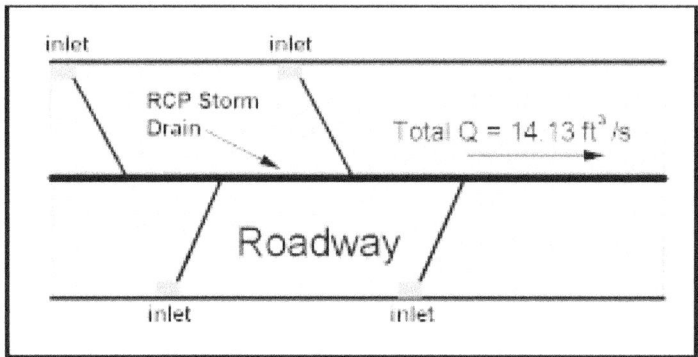

Find: The required storm drain pipe diameter and the full flow velocity.

1. Use the full flow equation (which gives pipe diameter in m)

$$Q = \frac{K_u}{n} D^{8/3} S^{1/2} \quad \text{where } K_u = 0.46$$

$$14.13 = \frac{0.46}{0.013} D^{8/3} (0.005)^{1/2}$$

$$D^{8/3} = 5.64$$

$$D = 1.91 \text{ ft}$$

2. Based on table 2, the next larger nominal pipe size is 24 in. or 2 ft.

3. Use the continuity equation (Q = VA) to calculate the full flow velocity.

Q = VA; therefore, V = Q/A

$$A = \frac{\pi D^2}{4}$$

$$= \frac{\pi (2.0)^2}{4} = 3.14 \text{ ft}^2$$

$$V = \frac{14.13}{3.14} = 4.50 \text{ ft/s}$$

4. Note: This is the full-flow velocity; however, under our design conditions, a 24 inch would be flowing slightly less than full. Based on the part-full flow relationships (figure 40) the velocity does not change significantly from half-full to full. However, for tc calculation, the part-full velocity should be used.

    To calculate the part-full velocity, nomographs or trial-and-error solution can be used. Alternatively, the part-full flow relationships can be used. The full-flow discharge and velocity of a 24-inch concrete pipe is

$$Q = \frac{0.46}{0.013} (2.0)^{8/3} (0.005)^{1/2} = 15.89 \text{ ft}^3 \text{/s}$$

$$V = \frac{Q}{A} = \frac{15.89}{3.14} = 5.06 \text{ ft/s}$$

The ratio of part-full to full-flow discharge is

$$\frac{Q}{Q_f} = \frac{14.13}{15.89} = .89$$

and from the part-full flow relationships (Figure 40), the corresponding velocity ratio is 1.12. Therefore

$$\frac{V}{V_f} = 1.12 \quad \text{and} \quad V = 1.12(5.06) = 5.66 \text{ ft/s}$$

6.3.1. Calculating Friction Losses

Friction losses are calculated as

$$h_f = LS_f \tag{52}$$

where:

L	=	Length of the conduit
S_f	=	Friction slope (energy grade line slope)

Uniform flow conditions are typically assumed so that the friction slope can be calculated from either Manning's equation, or the Darcy-Weisbach equation. Rewriting Manning's equation for S_f:

$$S_f = \left(\frac{Qn}{K_u AR^{2/3}} \right)^2 \tag{53}$$

The Darcy-Weisbach equation for open-channel flow:

$$S_f = \frac{f}{4R} \frac{V^2}{2g} \tag{54}$$

and for pressure flow in circular conduit:

$$h_f = \frac{fL}{D} \frac{V^2}{2g} \tag{55}$$

Manning's equation is more commonly used by practicing engineers, even though the Darcy-Weisbach equation is a theoretically better equation since it is dimensionally correct and applicable for any fluid over a wide range of conditions. However, the possibilities for greater accuracy with the Darcy-Weisbach equation are limited by determination of the Darcy f and a generally more complicated application than the Manning's equation. Typical Manning's n values for closed-conduit flow are given in table 15.

No matter which formula is used, judgment is required in selecting roughness coefficients. Roughness coefficients are primarily defined by the type of pipe material; however, many other factors can modify the value based on pipe material. Other important factors include the type of joint used, poor alignment and grade due to settlement or lateral soil movement, sediment deposits and flow from laterals disturbing flow in the mainline.

6.3.2. Calculating Form Losses

Form losses occur when flow passes through structures such as access holes, junctions, bends, contractions, enlargements and transitions. These structures can cause major losses in both the energy grade line and the hydraulic grade line across the structure, and if not accounted for in design, the capacity of the conduit may be restricted.

Form losses may be evaluated by several methods. The simplest method is based on a coefficient times the velocity head, with different coefficients tabulated for access holes, bends, inlets, etc. The general form of the equation is:

$$h_L = K \frac{V^2}{2g} \tag{56}$$

An alternate method is based on the sum of four individual losses defined as a function of velocity: entrance/exit losses, velocity correction loss, bend loss and junction loss. Perhaps the most sophisticated approach is based on pressure and momentum concepts, specifically that the sum of all pressures acting at a junction must equal the sum of all momentums.

The HYDRA module of HYDRAIN uses several methods to calculate form losses.[10] Bend losses are based on a coefficient times velocity head, with the coefficient (K) being a

122

function of the angle and radius of the bend. In HYDRAIN, version 5.0, HYDRA access hole losses are calculated in a similar fashion with K defined as

$$K = K_o \times C_D \times C_d \times C_Q \times C_p \times C_B \qquad (57)$$

where:

K	=	Adjusted headloss coefficient
K_o	=	Initial headloss coefficient based on relative access hole size
C_D	=	Correction factor for pipe diameter
C_d	=	Correction factor for flow depth
C_Q	=	Correction factor for relative flow
C_B	=	Correction factor for benching
C_p	=	Correction factor for plunging flow

A discussion on form loss relationships in appendix B provides equations to calculate these correction factors.

A pipe junction is the connection of a lateral pipe to a larger trunk pipe without the use of an access hole structure. In HYDRAIN, version 5.0, HYDRA adjoining pipes are considered a pipe junction when two inflow pipes (a lateral and a trunk) enter the junction (for more than two pipes, the access hole loss equations are employed). The minor loss equation for a pipe junction is a form of the momentum equation:

$$H_j = \frac{Q_o \times V_o - Q_i \times V_i - Q_1 \times V_1 \times \cos\theta}{0.5 \times g \times (A_o + A_i)} + h_i - h_o \qquad (58)$$

where:

H_j	=	Junction headloss, m (ft)
Q_o, Q_i, Q_1	=	Outlet, inlet, and lateral flows, respectively, m³/s (ft³/s)
V_o, V_i, V_1	=	Outlet, inlet, and lateral velocities, respectively, m/s (ft/s)
h_o, h_i	=	Outlet and inlet velocity heads, m (ft)
A_o, A_i	=	Outlet and inlet cross-sectional areas, m² (ft²)
θ	=	Angle of lateral with respect to centerline of outlet pipe, degrees
g	=	Gravitational acceleration, 9.81 m/s² (32.2 ft/s²)

EXAMPLE PROBLEM 6.2 (SI Units)

Given: An RCP (n = 0.013) storm drain, 600 mm in diameter, carries 0.30 m³/s on a 0.001 slope. The storm drain is 301.5 m long, with a 1.5 m (diameter) access hole in the middle. The HGL at the outlet is 1.50 m above a datum defined by the pipe invert. Calculate the HGL and EGL profiles. What is the minimum cover hydraulically required over the storm drain?

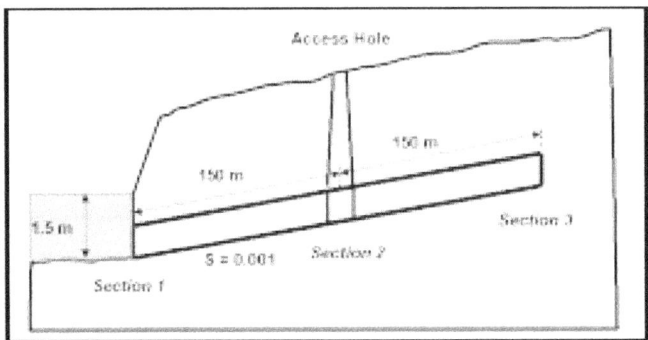

Find: HGL

EGL

Minimum burial depth

1. Check for pressure flow conditions: from the full-flow Manning's equation, the full flow capacity is

$$Q_{full} = \frac{K_u}{n} D^{8/3} S^{1/2} \quad \text{where } K_u = 0.312 \text{ for SI units}$$

$$Q_{full} = \frac{0.312}{0.013} (0.6)^{8/3} (0.001)^{1/2} = 0.2 \, m^3 / s$$

Since the design Q = 0.3 m³/s, pressure flow exists.

2. Flow velocity in pipe

$$Q = VA, \quad V = \frac{Q}{A}$$

$$A = \frac{\pi D^2}{4} = \frac{3.14 \, (0.6)^2}{4} = 0.28 \, m^2$$

$$V = \frac{0.30}{0.28} = 1.1 \, m/s$$

3. Calculate the losses and apply energy equation

The HGL at the outlet (section 1) was given as 1.5 using a datum at the invert of the pipe.

Assume:

$$Z_1 = 0$$

Then:
$$Z_2 = 150(0.001) = 0.15 \, m$$

Apply the energy equation from the ponded water to section 1, the outlet of the pipe

$$\frac{V_1^2}{2g} + \frac{P_1}{\gamma} + Z_1 = \frac{V_p^2}{2g} + \frac{P_p}{\gamma} + Z_p + h_L$$

The headloss is the exit loss created at the outlet

$$h_L = K \frac{V^2}{2g}$$

For the exit, the value of K is 1.0 and the loss is

$$h_L = 1.0 \frac{(1.1)^2}{2(9.81)} = 0.062$$

$$\frac{(1.1)^2}{2g} + \frac{P_1}{\gamma} + 0 = \frac{(0)^2}{2g} + 1.5 + 0 + 0.062$$

$$\frac{P_1}{\gamma} = 1.5 \, m$$

Apply the energy equation from section 1 to the downstream side of the access hole at section 2

$$\frac{V_{2d}^2}{2g} + \frac{P_{2d}}{\gamma} + Z_{2d} = \frac{V_1^2}{2g} + \frac{P_1}{\gamma} + Z_1 + h_r$$

The headloss is due to friction.

$$h_r = LS_f = L\left(\frac{Q_n}{K_u D^{8/3}}\right)^2 \quad \text{where } K_u = 0.312$$

Therefore, the friction headloss from section 1 to 2 and from section 2 to 3 is

$$h_r = 150\left[\frac{(0.3)\,(0.013)}{(0.312)\,(0.6)^{8/3}}\right]^2 = 0.37 \, m$$

$$\frac{(1.1)^2}{2(9.81)} + \frac{P_{2s}}{\gamma} + 0.15 = \frac{(1.1)^2}{2(9.81)} + 1.5 + 0 + 0.37$$

Therefore $\frac{P_{2s}}{\gamma} = 1.72$ m

Access Hole Loss

For the access hole, the coefficient K is defined as

$$K = K_o \times C_D \times C_d \times C_Q \times C_p \times C_B$$

As provided in HDS-4, appendix B, the initial headloss coefficient K_o, is estimated as a function of the relative access hole size and angle between the inflow and outflow pipes:

$$K_o = 0.1 \times [\frac{b}{D_o}] \times [1 - \sin\theta] + 1.4 \times [\frac{b}{D_o}]^{0.15} \times \sin\theta$$

where:

θ	=	Angle between the inflow and outflow pipes
b	=	Access hole diameter
D_o	=	Outlet pipe diameter

$$K_o = 0.1[\frac{1.5}{0.60}] \times [1 - \sin 180] + 1.4 \times [\frac{1.5}{0.60}]^{0.15} \times \sin 180$$

$$K_o = 0.25$$

The correction factor for pipe diameter (C_D) is 1.0 since the incoming and outgoing pipe diameters are the same. Similarly, the correction factor for relative flow (C_Q) is 1.0 since the inflow and outflow discharge rates are the same, the plunging flow correction (C_p) is 1.0 since there are not multiple inflows and with no benching $C_B = 1$. The correction factor for flow depth C_d is a function of the depth of water at the upstream end of the outlet pipe (1.72 m). Therefore,

$$C_d = 0.5 [\frac{d}{D_o}]^{3/5} = 0.5 [\frac{1.72}{0.6}]^{3/5} = 0.94$$

Therefore, the coefficient K for evaluating the access hole loss is

$$K = 0.25 \times 1.0 \times 0.94 \times 1.0 \times 1.0 \times 1.0 = 0.24$$

and the headloss is

$$h_L = K \frac{V^2}{2g} = 0.24 \left(\frac{(1.1)^2}{2 \times 9.81} \right) = 0.015 m$$

126

The headloss at a structure, such as an access hole, is typically assumed to be uniformly distributed across the structure. Apply the energy equation across the structure. Allow for a crown drop equal to the headloss through the access hole.

$$\frac{V_{2u}^2}{2g} + \frac{P_{2u}}{\gamma} + Z_{2u} = \frac{V_{2d}^2}{2g} + \frac{P_{2d}}{\gamma} + Z_{2d} + h_L$$

$$\frac{(1.1)^2}{2g} + \frac{P_{2u}}{\gamma} + (0.15 + 0.015) = \frac{(1.1)^2}{2g} + 1.72 + 0.15 + 0.015; \frac{P_{2u}}{\gamma} = 1.720$$

Then, applying the energy equation from section 2_u to section 3

$$\frac{V_3^2}{2g} + \frac{P_3}{\gamma} + Z_3 = \frac{V_{2u}^2}{2g} + \frac{P_{2u}}{\gamma} + Z_2 + h_f$$

$$\frac{P_3}{\gamma} + 0.315 = 1.720 + 0.165 + 0.37$$

$$\frac{P_3}{\gamma} = 1.940 \text{ m}$$

4.   Calculate the HGL and EGL at each section

$$HGL = \frac{P}{\gamma} + Z$$

$$EGL = HGL + \frac{V^2}{2g}$$

Section	P/γ	Z	HGL	V²/2g	EGL
1	1.50	0.00	1.50	0.06	1.56
2d	1.72	0.15	1.87	0.06	1.930
2u	1.72	0.165	1.885	0.06	1.945
3	1.94	0.315	2.255	0.06	2.315

5.   The minimum cover is that required to keep the HGL below the roadway grade, particularly at the access hole. The distance that the HGL is above the crown of the pipe at the access hole is

$$1.87 - 0.6 - 0.15 = 1.12 \text{ m} \quad \text{say } 1.2 \text{ m}$$

Therefore, the minimum cover is 1.2 m. Structural considerations may require greater cover.

EXAMPLE PROBLEM 6.2 (English Units)

Given: An RCP (n = 0.013) storm drain, 24 inches in diameter, carries 10.6 ft³/s on a 0.001 slope. The storm drain is 1005 ft long, with a 5 ft (diameter) access hole in the middle. The HGL at the outlet is 5 ft above a datum defined by the pipe invert. Calculate the HGL and EGL profiles. What is the minimum cover hydraulically required over the storm drain?

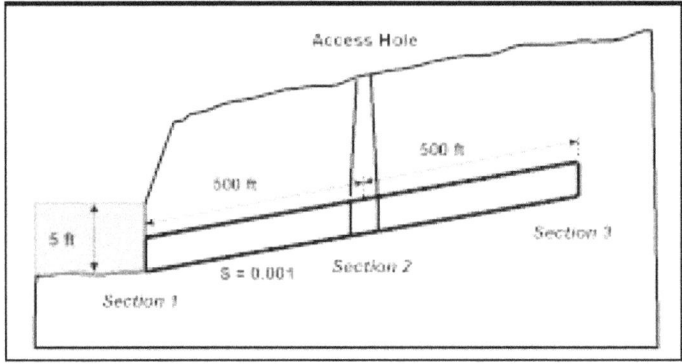

Find: HGL
　　　 EGL
　　　 Minimum burial depth

1.　　Check for pressure flow conditions: from the full-flow Manning's equation, the full flow capacity is

$$Q_{full} = \frac{K_u}{n} D^{8/3} S^{1/2} \quad \text{where } K_u = 0.46 \text{ for English units}$$

$$Q_{full} = \frac{0.46}{0.013} (2.0)^{8/3} (0.001)^{1/2} = 7.11 \, ft^3 / s$$

Since the design Q = 10.6 ft³/s, pressure flow exists.

2.　　Flow velocity in pipe

$$Q = VA, \, V = \frac{Q}{A}$$

$$A = \frac{\pi D^2}{4} = \frac{3.14 (2.0)^2}{4} = 3.14 \, ft^2$$

$$V = \frac{10.6}{3.14} = 3.38 \, ft/s$$

3.　　Calculate the losses and apply energy equation

The HGL at the outlet (section 1) was given as 5.0 using a datum at the invert of the pipe.

Assume:

$$Z_1 = 0$$

Then:

$$Z_2 = 500(0.001) = 0.5 \text{ ft}$$

Apply the energy equation from the ponded water to section 1, the outlet of the pipe

$$\frac{V_1^2}{2g} + \frac{P_1}{\gamma} + Z_1 = \frac{V_p^2}{2g} + \frac{P_p}{\gamma} + Z_p + h_L$$

The headloss is the exit loss created at the outlet

$$h_L = K\frac{V^2}{2g}$$

For the exit, the value of K is 1.0 and the loss is

$$h_L = (1.0)\frac{(3.38)^2}{2(32.2)} = 0.18 \text{ ft}$$

$$\frac{(3.38)^2}{2(32.2)} + \frac{P_1}{\gamma} + 0 = \frac{(0)^2}{2g} + 5.0 + 0 + 0.18$$

$$\frac{P_1}{\gamma} = 5.0 \text{ ft}$$

Apply the energy equation from section 1 to the downstream side of the access hole at section 2

$$\frac{V_{2d}^2}{2g} + \frac{P_{2d}}{\gamma} + Z_{2d} = \frac{V_1^2}{2g} + \frac{P_1}{\gamma} + Z_1 + h_f$$

The headloss is due to friction.

$$h_f = LS_f = L\left(\frac{Qn}{K_u D^{8/3}}\right)^2 \quad \text{where } K_u = 0.46$$

Therefore, the friction headloss from section 1 to 2 and from section 2 to 3 is

$$h_f = 500\left[\frac{(10.6)\,(0.013)}{0.46\,(2.0)^{8/3}}\right]^2 = 1.11 \text{ ft}$$

$$\frac{(3.38)^2}{2(32.2)} + \frac{P_{2d}}{\gamma} + 0.5 = \frac{(3.38)^2}{2(32.2)} + 5.0 + 0 + 1.11$$

Therefore $\frac{P_{2d}}{\gamma} = 5.61$ ft

Access Hole Loss

For the access hole, the coefficient K is defined as

$$K = K_o \times C_D \times C_d \times C_Q \times C_p \times C_B$$

As provided in HDS-4, appendix B, the initial headloss coefficient K_o, is estimated as a function of the relative access hole size and angle between the inflow and outflow pipes:

$$K_o = 0.1 \times [\frac{b}{D_o}] \times [1 - \sin\theta] + 1.4 \times [\frac{b}{D_o}]^{0.15} \times \sin\theta$$

where:

θ	=	Angle between the inflow and outflow pipes
b	=	Access hole diameter
D_o	=	Outlet pipe diameter

$$K_o = 0.1 \left[\frac{5.0}{2.0}\right] \times [1 - \sin 180] + 1.4 \times \left[\frac{5.0}{2.0}\right]^{0.15} \times \sin 180$$

$$K_o = 0.25$$

The correction factor for pipe diameter (C_D) is 1.0 since the incoming and outgoing pipe diameters are the same. Similarly, the correction factor for relative flow (C_Q) is 1.0 since the inflow and outflow discharge rates are the same, the plunging flow correction (C_p) is 1.0. Since there are not multiple inflows and with no benching, $C_B = 1$. The correction factor for flow depth C_d is a function of the depth of water at the upstream end of the outlet pipe. Therefore,

$$C_d = 0.5 \left[\frac{d}{D_o}\right]^{3/5} = 0.5 \left[\frac{5.61}{2.0}\right]^{3/5} = 0.93$$

Therefore, the coefficient K for evaluating the access hole loss is

$$K = 0.25 \times 1.0 \times 0.93 \times 1.0 \times 1.0 \times 1.0 = 0.23$$

and the headloss is

$$h_L = K\frac{V^2}{2g} = 0.23 \left(\frac{(3.38)^2}{2 \times 32.2}\right) = .04 \text{ ft}$$

The headloss at a structure, such as an access hole, is typically assumed to be uniformly distributed across the structure. Apply the energy equation across the structure. Allow for a crown drop equal to the headloss through the access hole.

$$\frac{V_{2_u}^2}{2g} + \frac{P_{2_u}}{\gamma} + Z_{2_u} = \frac{V_{2_d}^2}{2g} + \frac{P_{2_d}}{\gamma} + Z_{2_d} + h_L$$

$$\frac{(3.38)^2}{2g} + \frac{P_{2_u}}{\gamma} + (0.5 + 0.04) = \frac{(3.38)^2}{2g} + 5.61 + 0.5 + .04; \quad \frac{P_{2_u}}{\gamma} = 5.61$$

Then, applying the energy equation from section 2_u to section 3

$$\frac{V_3^2}{2g} + \frac{P_3}{\gamma} + Z_3 = \frac{V_{2_u}^2}{2g} + \frac{P_{2_u}}{\gamma} + Z_2 + h_f$$

$$\frac{(3.38)^2}{2g} + \frac{P_3}{\gamma} + 1.04 = \frac{(3.38)^2}{2g} + 5.61 + 0.94 + 1.11$$

$$\frac{P_3}{\gamma} + 1.04 = 5.61 + 0.54 + 1.11$$

$$\frac{P_3}{\gamma} = 6.22 \text{ ft}$$

4. Calculate the HGL and EGL at each section

$$HGL = \frac{P}{\gamma} + Z$$

$$EGL = HGL + \frac{V^2}{2g}$$

Section	P/γ	Z	HGL	V²/2g	EGL
1	5.0	0.00	5.0	0.18	5.18
2d	5.61	0.5	6.11	0.18	6.29
2u	5.61	0.54	6.15	0.18	6.33
3	6.22	1.04	7.26	0.18	7.44

5. The minimum cover is that required to keep the HGL below the roadway grade, particularly at the access hole. The distance that the HGL is above the crown of the pipe at the access hole is

    $6.11 - 2.0 - .5 = 3.61$ ft   say 3.7 ft

Therefore, the minimum cover is 3.7 ft. Structural considerations may require greater cover.

(page intentionally left blank)

7. APPLICATIONS OF CLOSED-CONDUIT FLOW

7.1. General Design Concepts

Typical closed-conduit facilities in highway drainage include culverts and storm drains. A storm drain facility can be a much more extensive closed-conduit system than a cross drainage system such as a culvert. In some respects, a storm drain is simply a long culvert. Storm drain systems consist of inlets connected to an underground pipe and an outlet facility. Storm drain systems are often used when the capacity of the roadway (established by the allowable spread) is exceeded, or for the collection and diversion of median drainage when the capacity of the swale is exceeded. A storm drain system may also be used in high gradient situations where erosion control is a concern.

Culverts are commonly used for cross drainage and can range in size from a single small culvert draining an isolated depression to multiple barrel designs and/or very large culverts for passing major stream channels under a roadway. Small culverts are also used for downdrains to protect fill slopes or to divert roadway water from a bridge deck.

Typical pipe materials used in storm drains include reinforced concrete pipe (RCP), corrugated metal pipe (CMP) and plastic pipe. These same materials are common for culverts, however, culverts are available in a variety of cross section shapes, and often a shape other than circular is desirable.

Conduit and culvert material are typically available in standard (nominal) sizes. Conduit size should not be decreased in the downstream direction, even if hydraulic calculations suggest this is possible. Energy dissipation is often required at the outlet of a storm drain or culvert to prevent erosion. Maintenance is required for any closed-conduit facility. Sediment deposition within the conduit and debris removal at the entrances are typical maintenance items.

7.2. Culvert Design

7.2.1. Design Approach

A culvert is a conduit that conveys flow through a roadway embankment or past some other type of flow obstruction. Culverts are typically constructed of concrete (reinforced and nonreinforced), corrugated metal (aluminum or steel) and plastic in a variety of cross sectional shapes. The most common cross sectional shapes for culverts are illustrated in figure 41a and typical entrance conditions are shown in figure 41b. The selection of culvert material depends on structural strength, hydraulic roughness, durability, and corrosion and abrasion resistance.

Flow conditions in a culvert may occur as open-channel flow, gravity full flow or pressure flow, or in some combination of these conditions. A complete theoretical analysis of the hydraulics of culvert flow is time-consuming and difficult. Flow conditions depend on a complex interaction of a variety of factors created by upstream and downstream conditions, barrel characteristics and inlet geometry. For purposes of design, standard procedures and nomographs have been developed to simplify the analysis of culvert flow. These procedures are detailed in Hydraulic Design Series Number 5 (HDS-5) entitled "Hydraulic Design of Highway Culverts."[27] The following information summarizes the basic design concepts and principles for culverts.

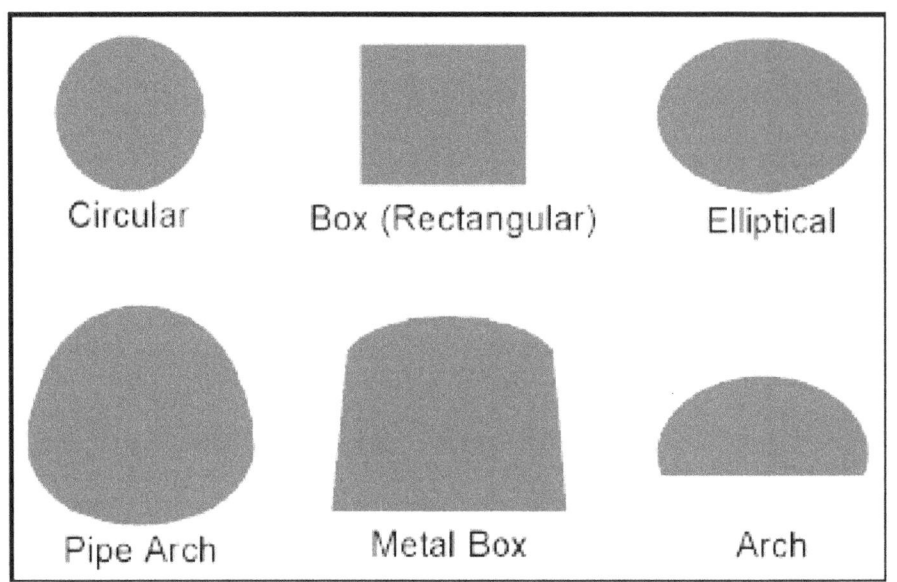

Figure 41a. Commonly used culvert shapes.

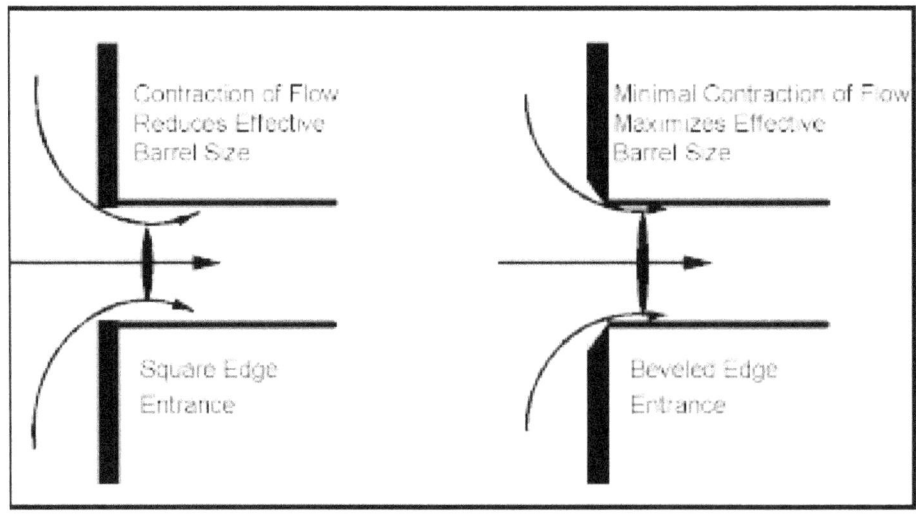

Figure 41b. Entrance contraction (schematic).

134

7.2.2. Types of Culvert Inlets and Outlets

A culvert typically represents a significant contraction of flow over conditions in the upstream and downstream channels, and often is a hydraulic control point in the channel. The provision of a more gradual flow transition at the inlet of a culvert can improve the discharge capacity of the culvert by reducing the energy losses associated with flow contraction. Culvert inlets are available in a variety of configurations and may be prefabricated or constructed in place. Commonly used inlet configurations include projecting culvert barrels, cast-in-place concrete headwalls, precast or prefabricated end sections, and culvert ends mitered to conform to the fill slope (figure 42). Structural stability, aesthetics, erosion control, fill retention, economics, safety, and hydraulic performance are considerations in the selection of an inlet.

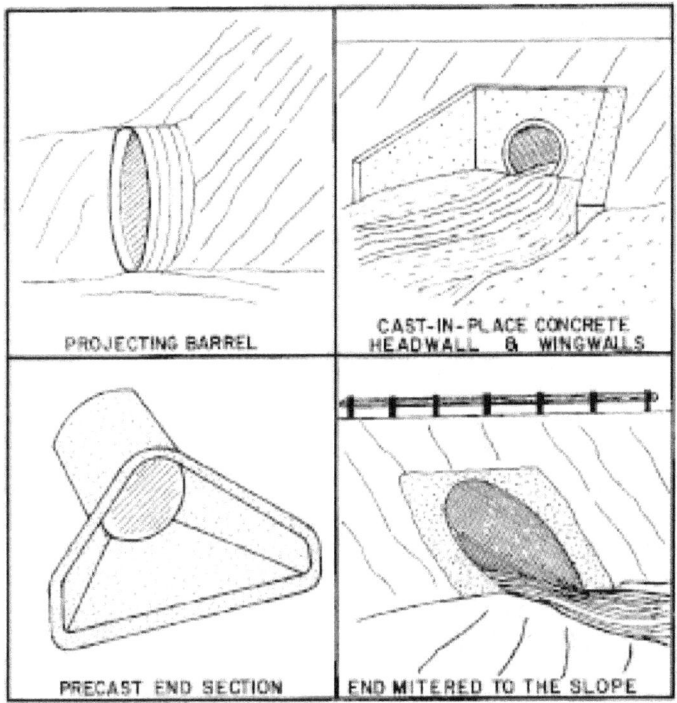

Figure 42. Four standard inlet types (schematic).

Hydraulic performance is improved by use of beveled edges rather than square edges, as illustrated in figure 41b. Side-tapered and slope-tapered inlets, commonly referred to as improved inlets, can significantly increase culvert capacity. Figure 43 illustrates side-tapered and slope-tapered inlet conditions. A side-tapered inlet provides a more gradual contraction of flow and reduces energy losses. A slope-tapered inlet, or depressed inlet, increases the effective head on the control section and improves culvert efficiency.

Culvert outlet configuration can be similar to any of the typical inlet configurations; however, hydraulic performance of a culvert is influenced more by tailwater conditions in the downstream channel than by the type of outlet. Outlet design is important for transitioning flow back into the natural channel, since outlet velocities are typically high and can cause scour of the downstream streambed and bank.

135

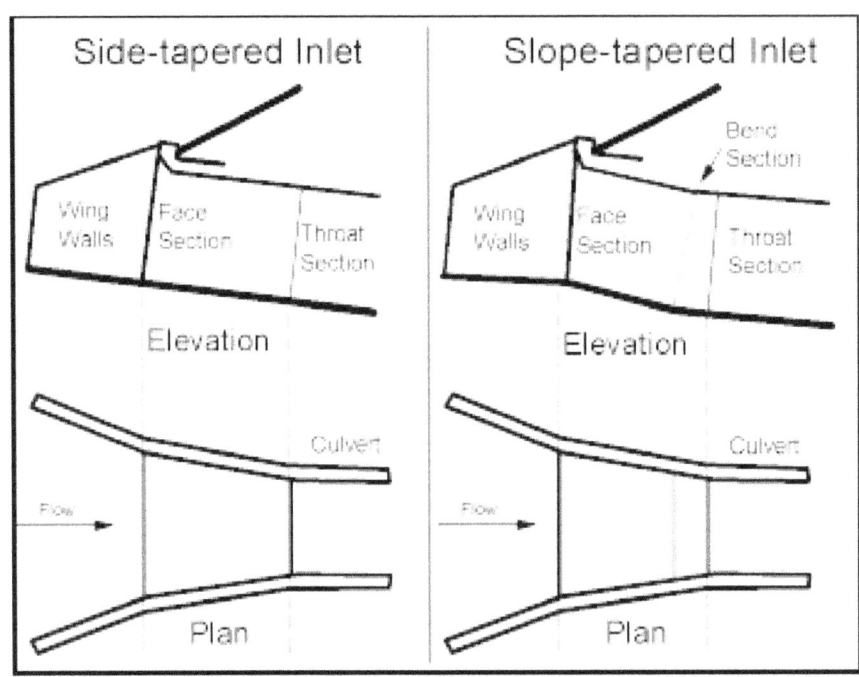

Figure 43. Side- and slope-tapered inlets.

7.2.3. Culvert Flow Conditions

A culvert may flow full over all its length or partially full. Full flow throughout a culvert is rare, and generally some portion of the barrel flows partly full. A water surface profile analysis is the only way to accurately determine how much of the barrel flows full. Pressure flow conditions in a culvert can be created by either high downstream or upstream water surface elevations. Regardless of the cause, the capacity of a culvert operating under pressure flow is affected by up- and downstream conditions and by the hydraulic characteristics of the culvert.

Partly full flow, or open-channel flow, in a culvert may occur as subcritical, critical or supercritical flow. Gravity full flow, where the pipe flows full with no pressure and the water surface just touches the crown of the pipe, is a special case of free surface flow and is analyzed in the same manner as open-channel flow.

7.2.4. Types of Flow Control

Based on a variety of laboratory tests and field experience, two basic types of flow control have been defined for culverts: (1) inlet control, and (2) outlet control. Inlet control occurs when the culvert barrel is capable of conveying more flow than the inlet will accept. The hydraulic control section of a culvert operating under inlet control is located just inside the entrance. Critical depth occurs at or near this location and the flow regime immediately downstream is supercritical. Hydraulic characteristics downstream of the inlet do not affect culvert capacity. The upstream water surface elevation and inlet geometry are the primary factors influencing culvert capacity.

136

Figure 44 illustrates typical inlet control conditions. The type of flow depends on the submergence of the inlet and outlet ends of the culvert; however, in each case the control section is at the inlet end of the culvert. For low headwater conditions the entrance of the culvert operates as a weir, and for headwaters submerging the entrance the entrance operates as an orifice. Figure 44a depicts a condition where neither the inlet nor the outlet end of the culvert are submerged. The flow passes through critical depth just downstream of the culvert entrance and the flow in the barrel is supercritical. The barrel flows partly full over its length, and the flow approaches normal depth at the outlet end.

Figure 44b shows that submergence of the outlet end of the culvert does not assure outlet control. In this case, the flow just downstream of the inlet is supercritical and a hydraulic jump forms in the culvert barrel.

Figure 44c is a more typical design situation. The inlet end is submerged and the outlet end flows freely. Again, the flow is supercritical and the barrel flows partly full over its length. Critical depth is located just downstream of the culvert entrance, and the flow is approaching normal depth at the downstream end of the culvert.

Figure 44d is an unusual condition illustrating the fact that even submergence of both the inlet and the outlet ends of the culvert does not assure full flow. In this case, a hydraulic jump will form in the barrel. The median inlet provides ventilation of the culvert barrel. If the barrel were not ventilated, sub-atmospheric pressures could develop which might create an unstable condition during which the barrel would alternate between full flow and partly full flow.

Outlet control occurs when the culvert barrel is not capable of conveying as much flow as the inlet opening will accept. The control section for outlet control is located at the barrel exit or further downstream. Either subcritical or pressure flow exists in the culvert under outlet control. All the geometric and hydraulic characteristics of the culvert play a role in determining culvert capacity. Figure 45 illustrates typical outlet control conditions. Condition 45a represents the classic full flow condition, with both inlet and outlet submerged. The barrel is in pressure flow throughout its length.

Condition 45b depicts the outlet submerged with the inlet unsubmerged. For this case, the headwater is shallow so that the inlet crown is exposed as the flow contracts into the culvert.

Condition 45c shows the entrance submerged to such a degree that the culvert flows full throughout its entire length while the exit is unsubmerged. This is a rare condition. It requires an extremely high headwater to maintain full barrel flow with no tailwater. The outlet velocities are unusually high under this condition.

Condition 45d is more typical. The culvert entrance is submerged by the headwater and the outlet end flows freely with a low tailwater. For this condition, the barrel flows partly full over at least part of its length (subcritical flow) and the flow passes through critical depth just upstream of the outlet.

Condition 45e is also typical, with neither the inlet nor the outlet end of the culvert submerged. The barrel flows partly full over its entire length, and the flow profile is subcritical.

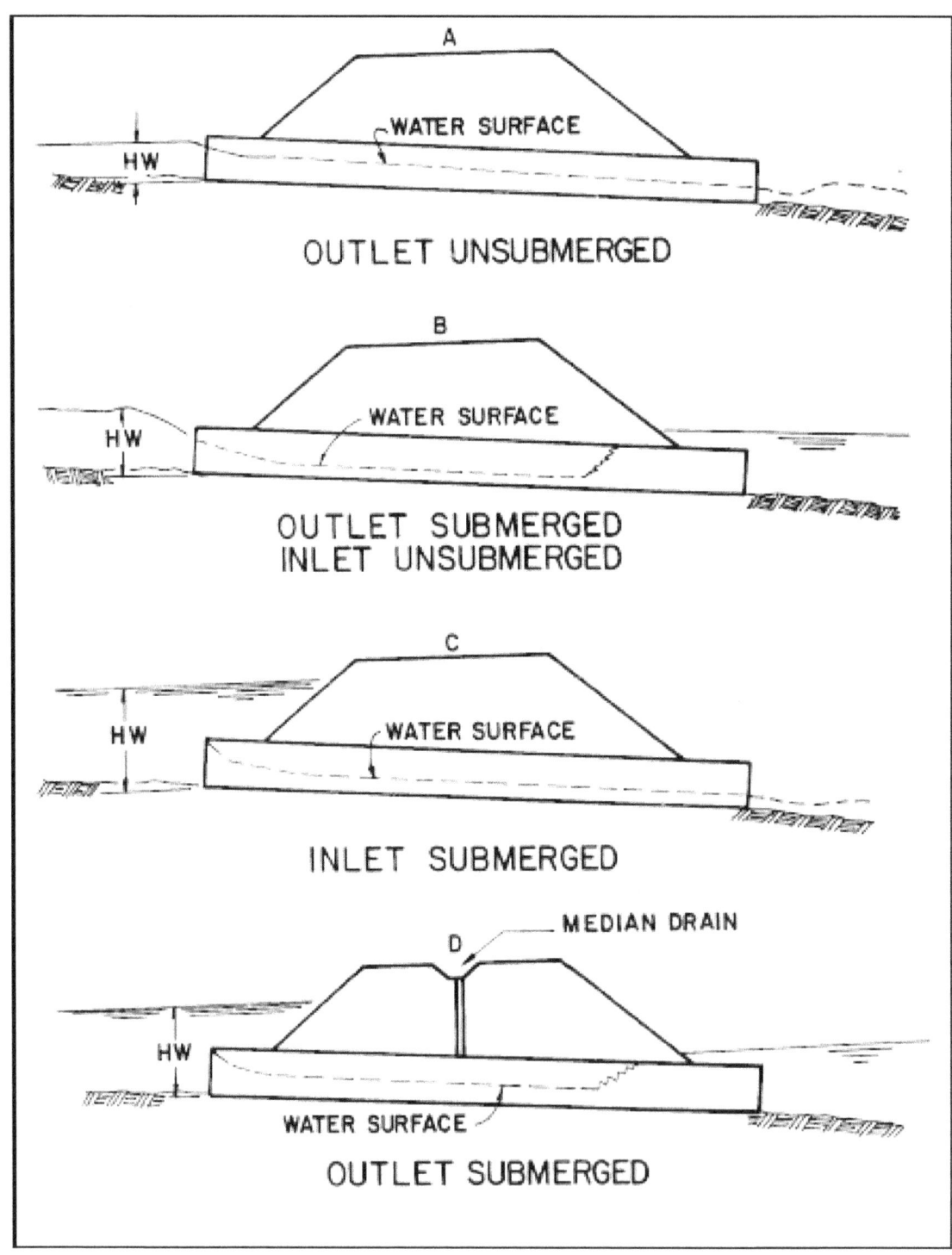

Figure 44. Types of inlet control

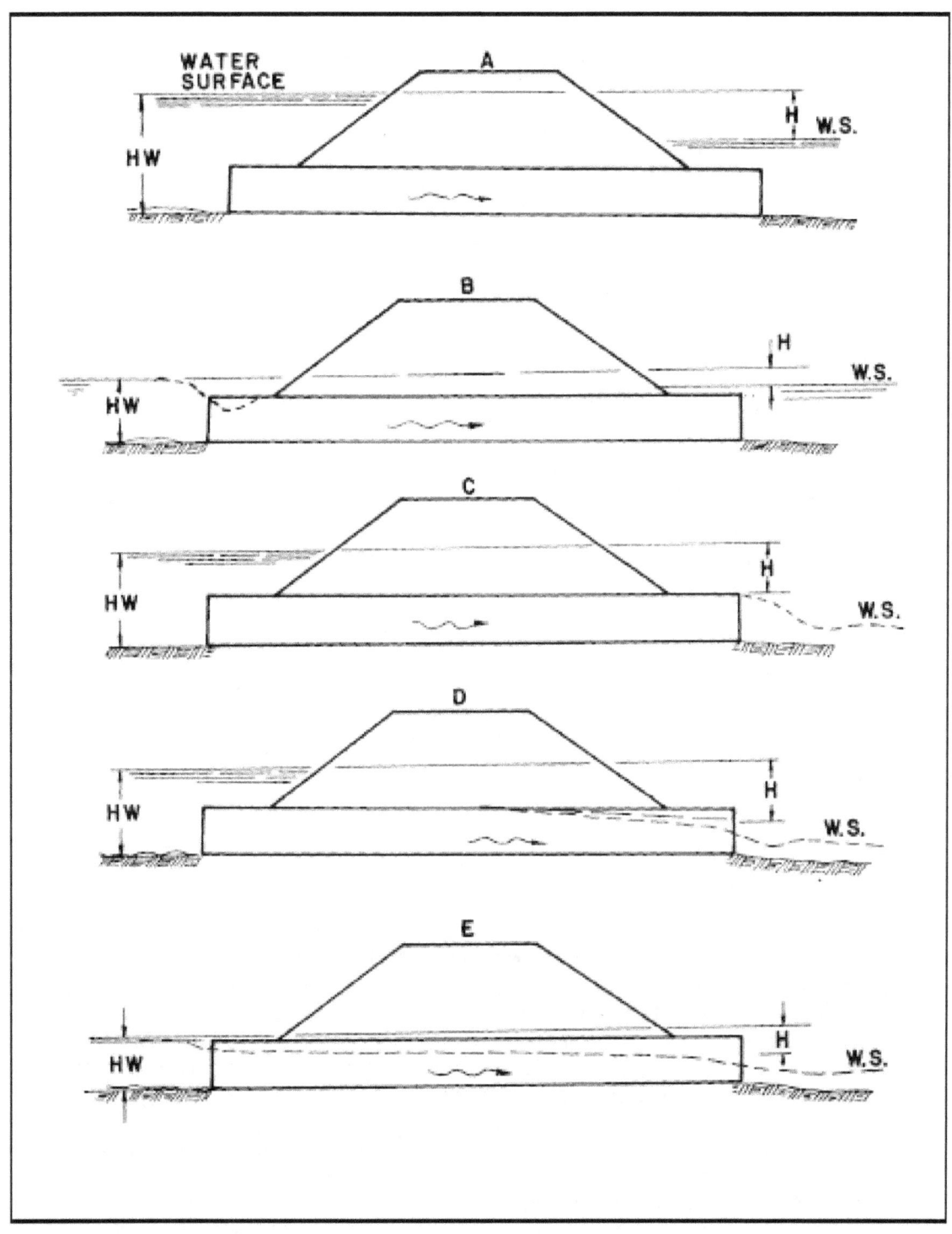

Figure 45. Types of outlet control.

7.2.5. Headwater and Tailwater Considerations

Energy is required to force flow through the constricted opening represented by a culvert. This energy occurs as an increased water surface elevation on the upstream side of the culvert. The headwater depth (HW) is defined as the depth of water at culvert entrance. In areas with flat ground slope or high fills a considerable amount of ponding may occur upstream of culvert. If significant, this ponding can attenuate flood peaks and may justify a reduction in the required culvert size.

Tailwater is defined as the depth of water downstream of the culvert, measured from the outlet invert. Tailwater is an important factor in determining culvert capacity under outlet control conditions. Tailwater conditions are most accurately estimated by water surface profile analysis of the downstream channel; however, when appropriate, tailwater conditions may be estimated by normal depth approximations.

7.2.6. Performance Curves

A performance curve is a plot of headwater depth or elevation versus flow rate. A performance curve can be used to evaluate the consequences of higher flow rates, such as the potential for overtopping the roadway if the design event is exceeded, or to evaluate the benefits of inlet improvements. In developing a performance curve both inlet and outlet control curves must be plotted, since the dominant control is hard to predict and may shift over a range of flow rates.

Figure 46 illustrates a typical culvert performance curve. Below elevation 4.3 headwater, the culvert operates under inlet control suggesting that inlet improvements might increase the culvert capacity and take better advantage of the culvert barrel capacity. A culvert that operates with inlet control over the range of design conditions could also be designed with additional barrel roughness to reduce outlet velocities, should downstream erosion be a concern.

7.2.7. Culvert Design Method

The basic design method is based on the location of the control (inlet or outlet). Although control may oscillate from inlet to outlet, the concept of "minimum performance" is applied meaning that while the culvert may operate more efficiently at times, it will never operate at a lower of performance than calculated. The design procedure then is to assume a pipe size and material and calculate the headwater elevation for both inlet and outlet control. The higher of the two is designated as the controlling headwater elevation. The controlling headwater elevation is compared to the desired design headwater, usually governed by overtopping considerations, to determine if the assumed culvert size is acceptable.

The outlet velocity should then be considered to evaluate the need for outlet protection. If the controlling headwater is based on inlet control, determine the normal depth and velocity in the culvert barrel. The velocity at normal depth is assumed to be the outlet velocity. If the controlling headwater is based on outlet control, determine the area of flow at the outlet based on the barrel geometry and the following: (1) critical depth if the tailwater is below critical depth, (2) the tailwater depth if the tailwater is between critical depth and the top of the barrel, (3) the height of the barrel if the tailwater is above the top of the barrel.

140

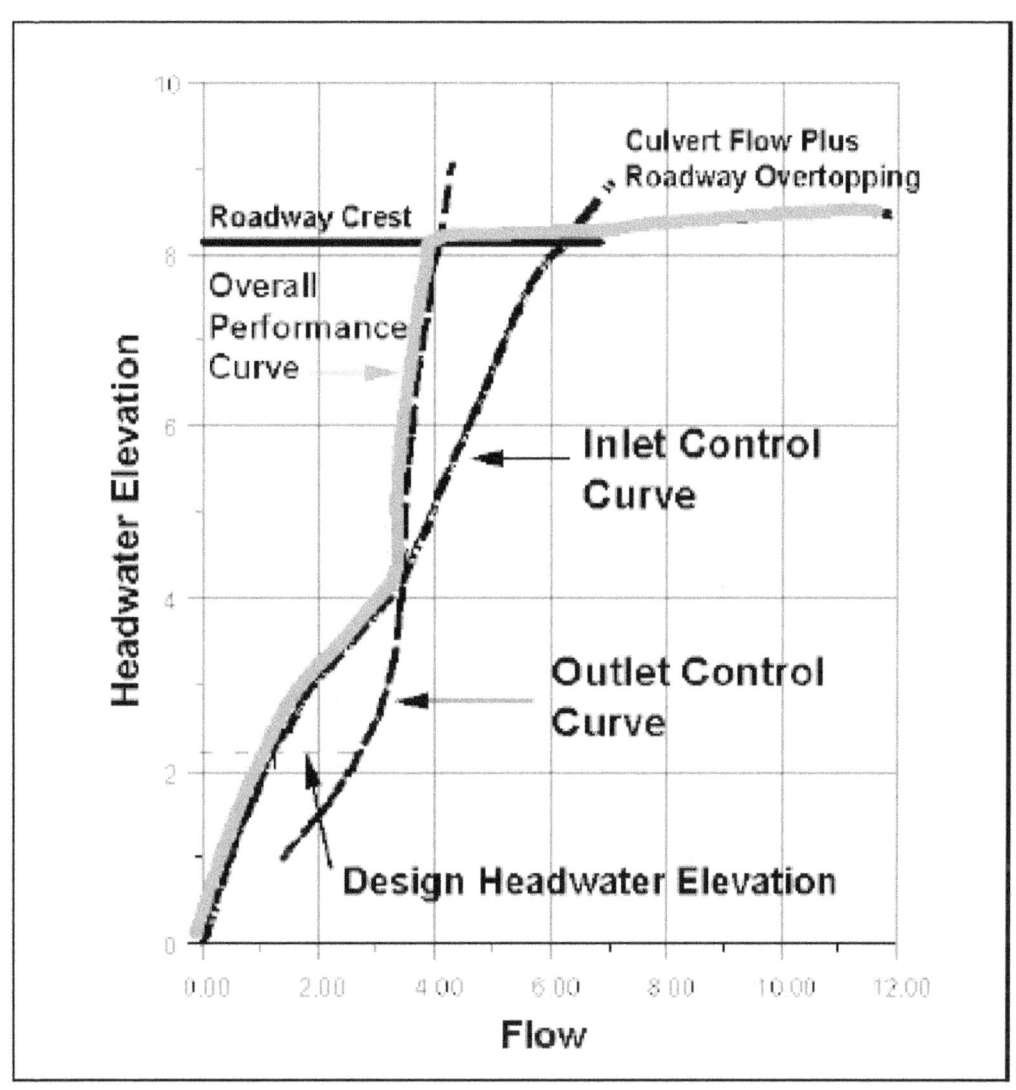

Figure 46. Culvert performance curve.

The evaluation of headwater conditions and outlet velocity is repeated until an acceptable culvert configuration is determined. To facilitate the design process, a Culvert Design Form is provided in HDS-5 and is reproduced in appendix B.[27] Once the barrel is selected, it must be fitted into the roadway cross section. The culvert barrel must have adequate cover, the length should be close to the approximate length, and the headwalls and wingwalls must be dimensioned.

An exact theoretical analysis of culvert flow is extremely complex because the flow is usually nonuniform with regions of both gradually varied and rapidly varied flow. An exact analysis would involve backwater and drawdown calculations, energy and momentum balance and applications of the results of hydraulic model studies. The flow conditions in a given culvert will change as the flow rate and tailwater elevations change, and hydraulic jumps often form inside or downstream of the barrel.

141

To avoid the analytical complexities created by this wide range of flow conditions, HDS-5 provides a culvert design method based on design charts and nomographs.[27] These same procedures are implemented by the computer program HY-8, available in HYDRAIN.[10] The design equations used to develop the nomograph and HY-8 procedures were based on extensive research. This research included quantifying empirical coefficients for various culvert conditions. Inlet and outlet control nomographs for RCP are provided in figures 47 and 48, respectively. HDS-5 provides nomographs for other pipe materials and shapes, and a number of examples on the application of the design method.[27] While it is possible to use the design method nomographs in HDS-5, and particularly the HY-8 computer program, without a thorough understanding of culvert hydraulics, this is not recommended.[10]

7.2.8. Improved Inlet Design

Culvert outlet control capacity is governed by headwater depth, tailwater depth, entrance configuration and barrel characteristics. The entrance condition is defined by the barrel cross-sectional area, shape and edge condition, while the barrel characteristics are area, shape, slope, length and roughness. Inlet improvements on culverts functioning under outlet control will reduce entrance losses, but these losses are only a small portion of the total headwater requirement. Therefore, only minor modifications of the inlet geometry which result in little additional cost are justified.

Culvert inlet control capacity is governed by only entrance configuration and headwater depth. Barrel characteristics and tailwater depth are normally of little consequence since culverts with inlet control typically flow only partly full. Entrance improvements can result in full, or nearly full flow, thereby increasing culvert capacity significantly.

As discussed in section 7.2.2, inlet improvements consist of bevel-edged inlets, side-tapered inlets and slope-tapered inlets. Beveled edges reduce the contraction of flow by effectively enlarging the face of the culvert.[27] Bevels are plane surfaces, but rounded edges that approximate a bevel and the socket end of RCP are also effective. Bevels are recommended on all headwalls.

A second degree of improvement is a side-tapered inlet. Tapered inlets improve culvert performance by providing a more efficient control section (the throat).[27] The inlet has an enlarged face area with the transition to the culvert barrel accomplished by tapering the sidewalls. The inlet face has the same height as the barrel, and its top and bottom are extensions of the top and bottom of the barrel. The intersection of the sidewall taper and barrel is defined as the throat section. Two control sections occur on a side-tapered inlet: at the face and throat. Throat control reduces the contraction at the throat.

A third degree of improvement is a slope-tapered inlet. The advantage of a slope-tapered inlet over a side-tapered inlet without a depression is that more head is applied at the control (throat) section. Both face and throat control are possible in a slope-tapered inlet; however, since the major cost of a culvert is in the barrel portion and not the inlet structure, the inlet face should be designed with greater capacity at the allowable headwater elevation than the throat. This will insure flow control will be at the throat and more of the potential capacity of the barrel will be used.

142

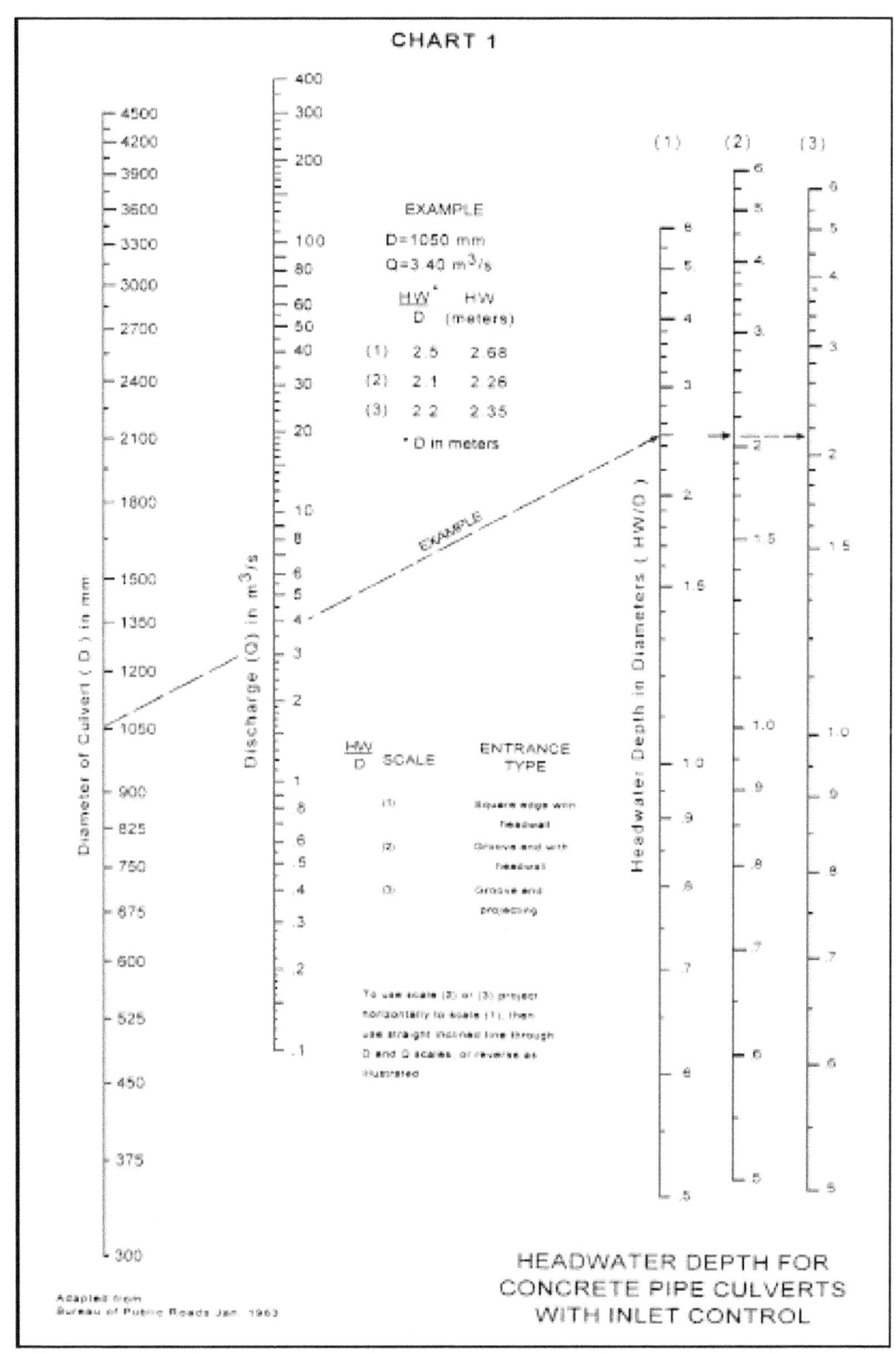

Figure 47a. RCP inlet control culvert nomograph - SI units.

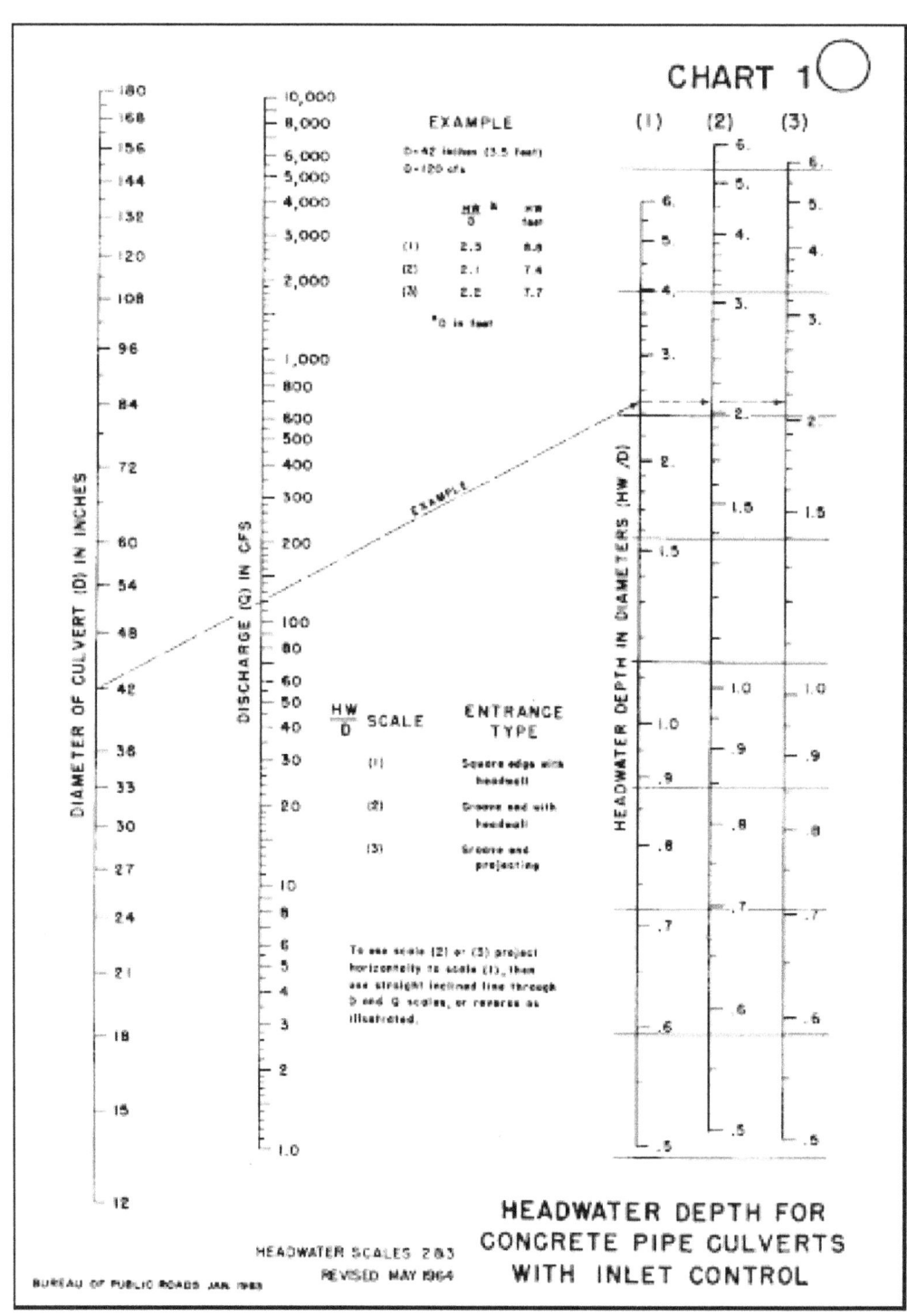

Figure 47b. RCP inlet control culvert nomograph - English units.

144

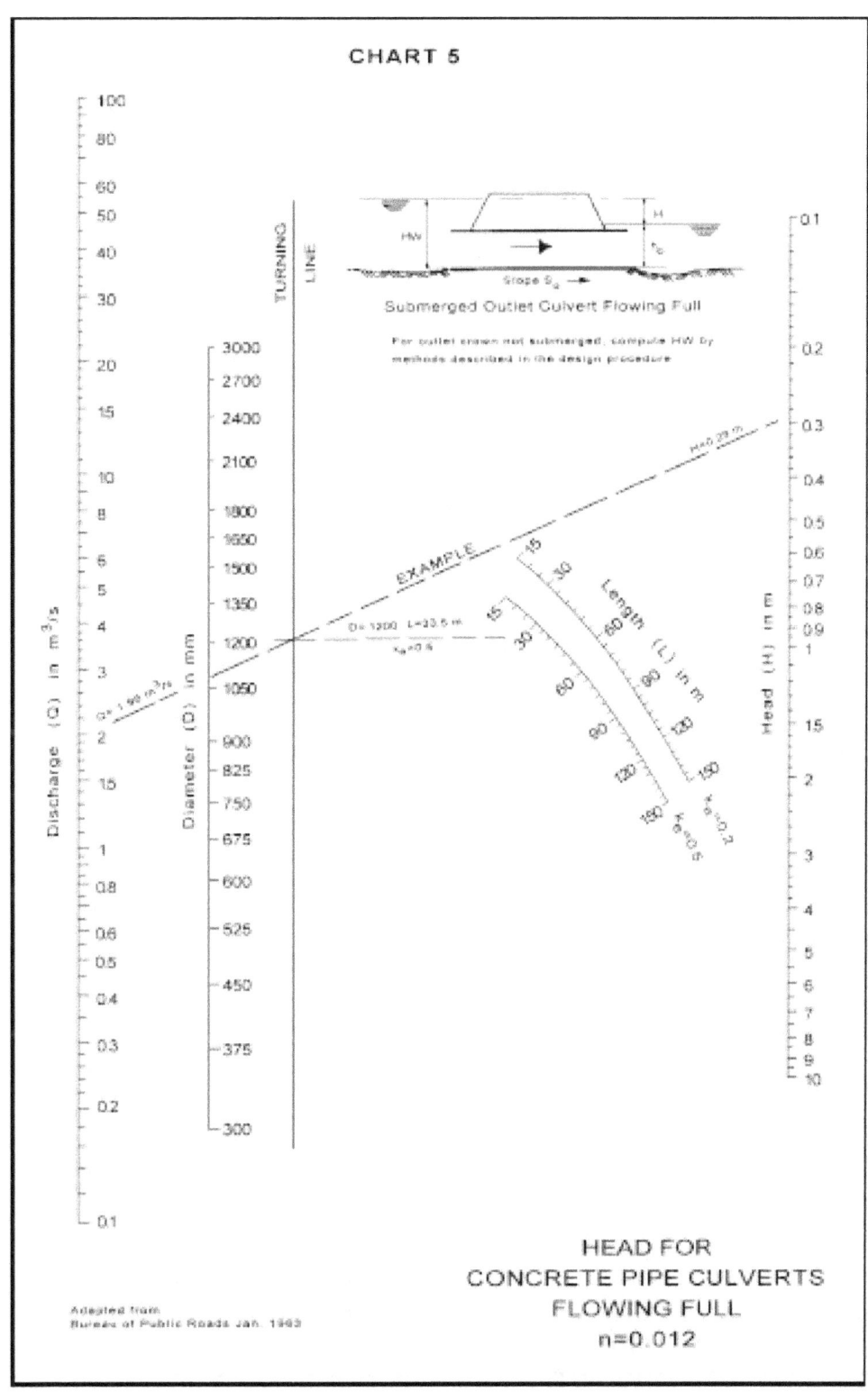

Figure 48a. RCP outlet control culvert nomograph - SI units.[27]

145

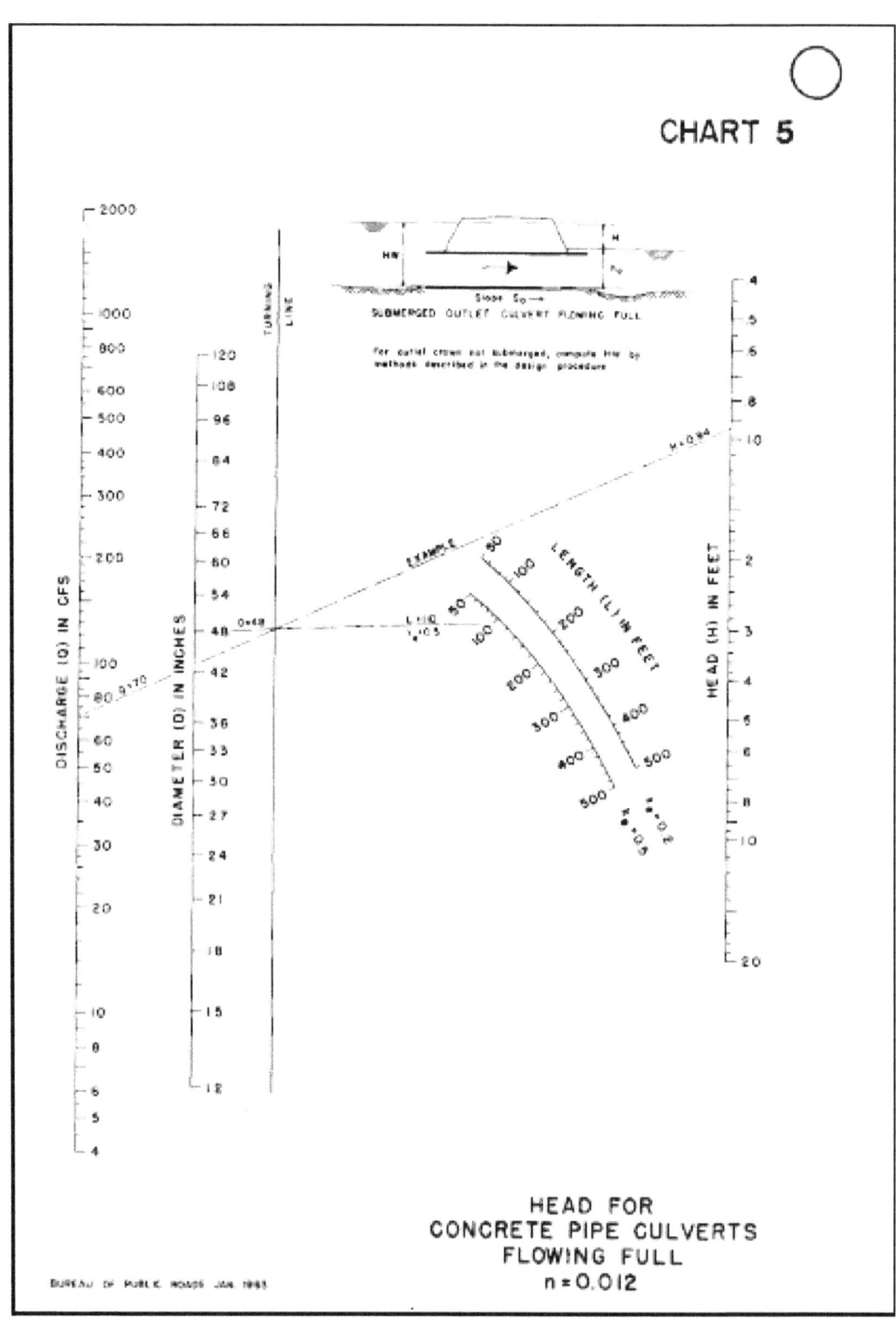

CHART 5

HEAD FOR
CONCRETE PIPE CULVERTS
FLOWING FULL
n = 0.012

Figure 48b. RCP outlet control culvert nomograph - English units.

EXAMPLE PROBLEM 7.1 (SI and English Units)

Given: A culvert at a new roadway crossing must be designed to pass the 25-year flood. Hydrologic analysis indicates a peak flow rate of 6.0 m^3/s (212 ft^3/s). The approximate culvert length is 60 m (197 ft) and the natural stream bed slope approaching the culvert is 1 percent. The elevation of the culvert inlet invert is 600 m (1968.5 ft) and the roadway elevation is 603 m (1978.35 ft). To provide some capacity in excess of the design flood, the desired headwater elevation should be at least 0.5 m (1.64 ft) below the roadway elevation. The tailwater for the 25-year flood is 1 m (3.28 ft).

Find: The size of RCP culvert necessary for the 25-year flood.

1. The design will be completed using the nomographs in figures 47 and 48. The Culvert Design Form will be used to facilitate the trial and error design process. The Culvert Design Form provides a summary of all the pertinent design data, and a small sketch with important dimensions and elevations.

2. The completed Culvert Design Form (see following page) indicates that a 1,500-mm (60 in.) RCP with a projecting groove end entrance, operating under inlet control, will result in a headwater elevation that is 0.9 m (2.95 ft) below the roadway.

3. The outlet velocity can be computed by calculating the full flow discharge (equation 51) and full flow velocity (from continuity), and then using the part-full flow relationships (figure 40) to find V/V$_f$ ratio given Q/Q$_f$. The computed outlet velocity is relatively high and protection should be provided at the outlet.

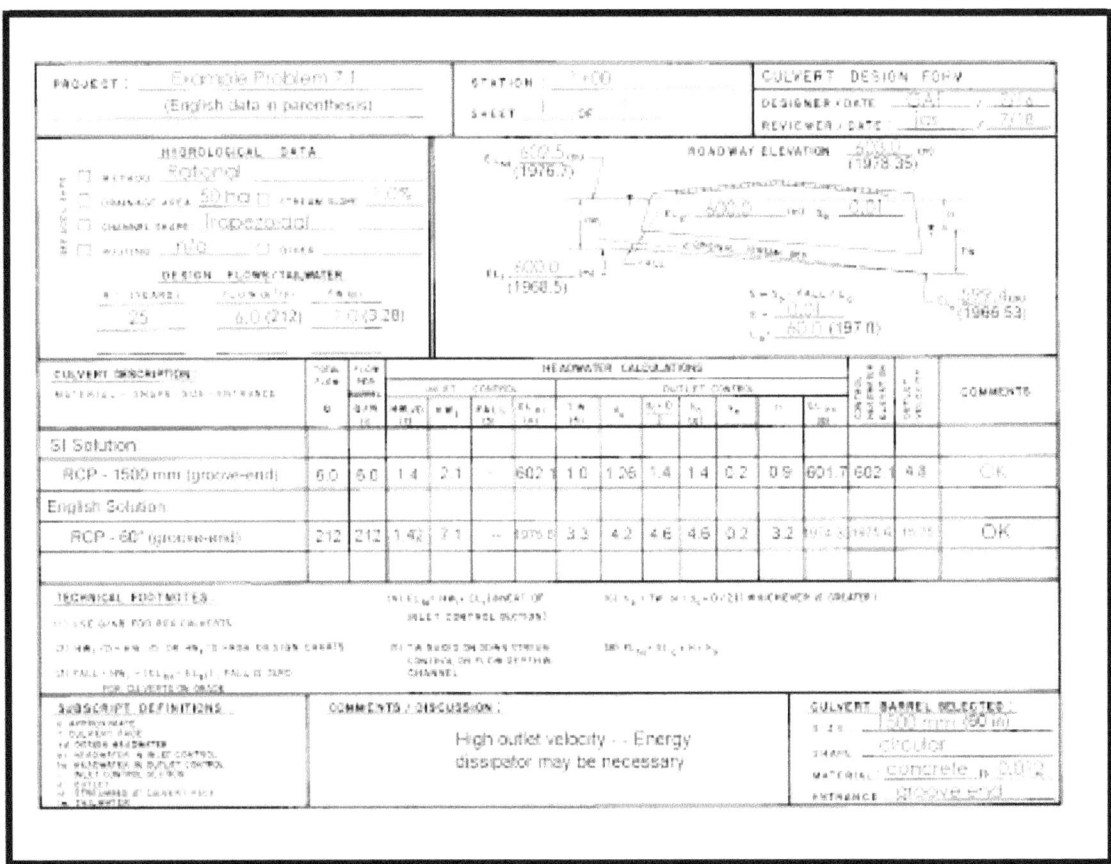

7.2.9. Culvert Design Using HYDRAIN

The culvert program in HYDRAIN is HY-8.[10] HY-8 is an interactive culvert analysis program that uses the HDS-5 analysis methods and information published by pipe manufacturers. The program will compute the culvert hydraulics and water surface profiles for circular, rectangular, elliptical, pipe arch, metal box and user-defined geometry. Additionally, improved inlets can be specified and the user can analyze inlet and outlet control for full and partially full culverts, analyze the tailwater in trapezoidal and coordinate defined downstream channels, analyze flow over the roadway embankment, and balance flows through multiple culverts.

7.3. Storm Drain Design

7.3.1. Design Approach

The design of a storm drain system is not a complicated process, but can involve detailed calculations that are often completed in an iterative manner. The major steps in storm drain design are:

1.   Development of a preliminary layout.

148

2.   Sizing the conduits in the storm drain system.

3.   Computation of the HGL.

4.   Adjustment of inlet sizes/locations and pipe size/location to correct HGL problems and/or optimize the design.

The following sections briefly discuss each major component in the design process.

7.3.2. Preliminary Layout

The first step in storm drain design is to develop a preliminary storm drain layout, including inlet, access hole and pipe locations. This is usually completed on a plan view map that shows the roadway, bridges, adjacent land use conditions, intersections, and under/overpasses. Other utility locations and situations should also be identified and shown, including surface utilities, underground utilities and any other storm drain systems. Storm drain alignment within the road right-of-way is usually influenced, if not dictated, by the location of other utilities. These other utilities, which may be public or private, may cause interference with the alignment or elevation of the proposed storm drain.

Generally, a storm drain should be kept as close to the surface as minimum cover and/or hydraulic requirements allow to minimize excavation costs. Another location control is the demand of traffic and the need to provide for traffic flow during construction including the possible use of detours. Providing curved storm drain alignments may be cost effective and should be considered for large pipe sizes, especially when headlosses are a concern. The deflection angle is divided by the allowable deflection per joint to determine the number of pipe sections required to create a given curve.

Tentative inlets, junctions and access locations should be identified based primarily on experience factors. The initially estimated type and location of inlets will provide the basis for hydrologic calculations and pipe sizing, and will be adjusted as required during the design process. Ultimately, inlets must be provided based on spread criteria and/or intersection requirements. Generally, all flow approaching an intersection should be intercepted, as cross gutters are not practical in highway applications.

Access is required for inspection and maintenance of storm drain systems. For storm drains smaller than about 1.2 m (48 in.), access is required about every 120 m (400 ft), while for larger sizes the spacing can be 180 m (600 ft) and larger. Junctions are also required at the confluence of two or more storm drains, where pipe size changes, at sharp curves or angle points (greater than 10°), or at abrupt grade changes.

7.3.3. Pipe Sizing

Given the preliminary layout, it is possible to begin the hydrologic and hydraulic analysis necessary to size the storm drain system. The first step is to calculate the discharge contributing to each inlet location, and to size the inlet. Based on the small incremental drainage areas involved, the Rational Method is typically used (chapter 2). Given the discharge at each selected inlet and considering spread criteria it may be necessary to relocate an inlet or to incorporate additional inlets.

After adjusting inlet locations the storm drain laterals and main line can be sized. Laterals are sized based on the discharge used to size the inlets. However, the discharge for the main line is not simply the sum of the incremental discharges at each inlet. Recalling that in the Rational Method the rainfall intensity used should be based on the longest time of concentration to the design point, the discharge for the main line should be determined based on longest time of concentration from various upstream approach branches and the corresponding accumulated CA values. This procedure satisfies the assumptions and stipulations for use of the Rational Method.

Preliminary pipe size is then calculated based on a full flow assumption given the discharge and pipe slope. This approach does not account for minor losses, which will be accounted for in the HGL calculation. Minor losses can be approximately accounted for at this stage of design by using a slightly higher roughness value in the full flow calculation. The pipe slope is typically established in preliminary design based on the roadway grade and the need to avoid other existing utilities or storm drains. When pipe sizes are increased in a downstream direction, it is generally preferable to match the crown elevation, rather than the invert elevation. The crown of the downstream pipe should drop by the headloss across the structure. Generally, storm drains should be designed to provide a velocity of at least 1 m/s (3 ft/s) when the conduit is full to insure that the pipe is self cleaning.

7.3.4. Computation of Hydraulic Grade Line

The HGL should be established to evaluate overall system performance and insure that at the design discharge the storm drain system does not inundate or adversely affect inlets, access holes or other appurtenances. The first step in calculating the HGL is establishing the location of all hydraulic controls along the conduit alignment and locations where the water surface must not be exceeded. These may occur at the inlet, at the outlet or at intermediate points along the alignment. For example, a median inlet will have an allowable maximum water surface elevation that prevents overtopping the median swale and flooding the roadway. The water surface elevation of the channel banks into which the storm drain discharges may establish the maximum tailwater. If the tailwater elevation is not known or is low, use the process given in HDS-5 to estimate the approximate hydraulic grade line by averaging the height of the storm drain and critical depth.[27] This value cannot exceed the height of the drain and should be used if it is greater than the actual tailwater depth. Storm drains are normally designed to flow full.

The calculation proceeds on a reach-by-reach basis from a given control point. In contrast with the downstream direction of calculations used to design the overall system, the calculation of the HGL usually proceeds in the upstream direction depending on the location of the control(s). The calculation is based on application of the energy equation (equation 4). For the first reach the energy losses (friction and form) are calculated. The first reach is defined from the downstream control point to the first hydraulic structure or conduit grade break. Subsequent reaches are defined between hydraulic structures and/or grade breaks. The design procedure is usually based on the assumption of a uniform hydraulic gradient within a conduit reach. Greater accuracy could be achieved with water surface profile computations, but such accuracy is seldom necessary.

EXAMPLE PROBLEM 7.2 (SI Units)

Given: A storm drain system is necessary to control erosion in the median of an interstate highway. The storm drain will discharge into a cross drainage culvert at a sag in the roadway profile. Design the preliminary pipe sizes for the main line of the storm drain system using corrugated metal pipe (CMP). Use n=0.024 for the CMP. Hydrologic analysis based on the Rational Method was completed for the inlets using the incremental drainage areas and time of concentration, and for the main line using the cumulated area and land use types (CA value) and the longest time of concentration. The resulting design discharge for line AB was 0.10 m³/s and 0.30 m³/s for line BC.

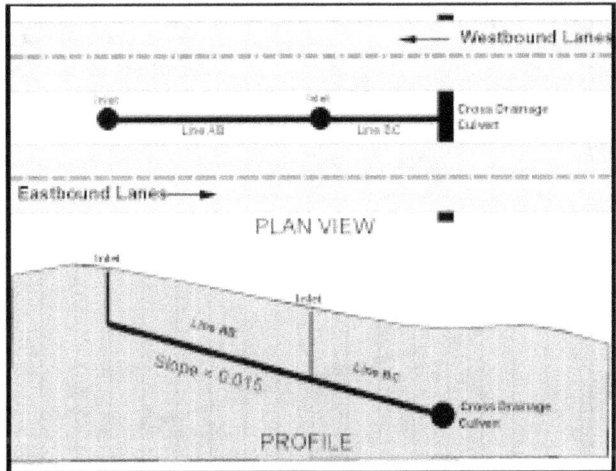

Find: Preliminary pipe sizes for the main line.

1. Line AB

$$Q = \frac{K_u}{n} D^{\frac{8}{3}} S^{\frac{1}{2}} \quad \text{where } K_u = 0.312 \text{ for SI Units}$$

$$D = \left[\frac{Qn}{0.312 S^{\frac{1}{2}}} \right]^{\frac{3}{8}} = \left[\frac{0.10(0.024)}{0.312(0.015)^{\frac{1}{2}}} \right]^{\frac{3}{8}} = 0.35 \text{ m}$$

Due to debris considerations and maintenance issues the minimum recommended pipe size for any storm drain system is 450 mm. Therefore, even though hydraulic analysis would suggest a smaller pipe size, use the recommended minimum pipe size of 450 mm for Line AB.

2. Line BC

$$D = \left[\frac{0.3(0.024)}{0.312(0.015)^{1/2}} \right]^{1/2} = 0.53 \text{ m (Use a 600mm nominal size)}$$

151

EXAMPLE PROBLEM 7.2 (English Units)

Given: A storm drain system is necessary to control erosion in the median of an interstate highway. The storm drain will discharge into a cross drainage culvert at a sag in the roadway profile. Design the preliminary pipe sizes for the main line of the storm drain system using corrugated metal pipe (CMP). Use n=0.024 for the CMP. Hydrologic analysis based on the Rational Method was completed for the inlets using the incremental drainage areas and time of concentration, and for the main line using the cumulated area and land use types (CA value) and the longest time of concentration. The resulting design discharge for line AB was 3.53 ft³/s and 10.59 ft³/s for line BC.

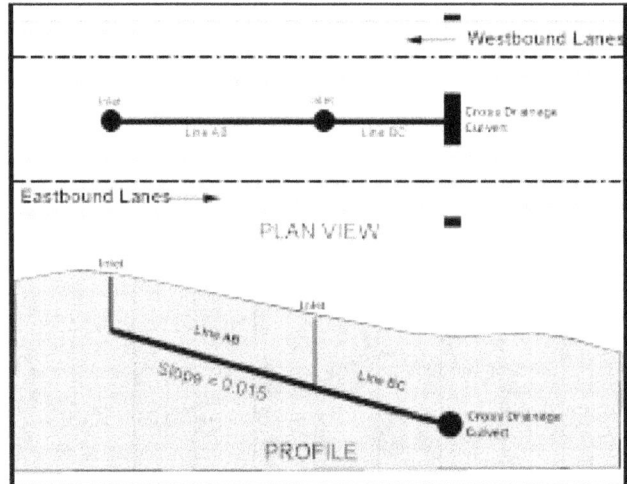

Find: Preliminary pipe sizes for the main line.

1. Line AB

$$Q = \frac{K_u}{n} D^{8/3} S^{1/2} \quad \text{where } K_u = 0.46 \text{ for English Units}$$

$$D = \left[\frac{Qn}{0.46\,S^{1/2}} \right]^{3/8} = \left[\frac{(3.53)(0.024)}{(0.46)(0.015)^{1/2}} \right]^{3/8} = 1.17 \text{ ft}$$

Due to debris considerations and maintenance issues the minimum recommended pipe size for any storm drain system is 18 inches. Therefore, even though hydraulic analysis would suggest a smaller pipe size, use the recommended minimum pipe size of 18 inches for Line AB.

2. Line BC

$$D = \left[\frac{(10.59)(0.024)}{(0.46)(0.015)^{1/2}} \right]^{3/8} = 1.76 \text{ ft (Use a 24 – inch nominal size)}$$

152

Friction losses are calculated assuming full flow. An appropriate equation for this calculation is the full flow version of the Manning's equation (equation 53), which defines the friction slope given the design discharge and preliminary pipe diameter. If the calculated friction slope is steeper than the pipe slope, pressure flow conditions will exist. If the friction slope is less than the pipe slope, partial flow will occur. At the location where the pipe becomes unsealed, or transitions to partial flow, normal depth calculations may be used to estimate hydraulic conditions. For each hydraulic structure, account for the form losses.

The energy losses are added to the energy grade line when working upstream, or subtracted when working downstream. The HGL is then calculated by subtracting the velocity head from the energy grade line, and the results typically plotted on the storm drain profile plan sheet. Form losses are typically assumed to occur as point losses across a given structure.

7.3.5. Optimization of System

Having completed the above calculations, initial design should be evaluated using a higher check flood and adjusted to reduce cost and risk. For example, if the HGL is too high in a given reach, the pipe size will have to be increased which will require recalculation of the HGL. As designed the system will operate basically at or near gravity full flow; however, if surcharging (pressure flow) is acceptable, the pipe sizes can be reduced and the system reanalyzed. The initial design should be evaluated using a higher check flood and adjusted to reduce cost and risk, if necessary.

7.3.6. Storm Drain Design Using HYDRAIN

Storm drain analysis and design can be completed in HYDRAIN, using the HYDRA module.[10] HYDRA generates storm flows by using either the Rational Method technique, hydrologic simulation techniques, or accepting a hydrograph generated by a HYDRO analysis. It can be used to design or analyze storm, sanitary or combined collection systems. HYDRA can handle up to 1,000 contributing drainage areas and 2,000 pipes. Additionally, HYDRA can be used for cost estimating. The Rational Method approximates the peak rate of runoff from a basin resulting from storms of a given return period. HYDRA's hydrologic simulation models the natural rainfall-runoff process. In the simulation, runoff hydrographs are generated, merged together, and routed through the collection system. Inlet limitations can be analyzed, inlet overflow can be passed down a gutter system, while inlets in sumps can store water in ponds.

In the HYDRA design process, the program will select the pipe size, slope, and invert elevations given certain design criteria. Additionally, HYDRA will perform analyses on an existing system of pipes (and/or ditches). When an existing system of pipes is overloaded, HYDRA will show suggested flow removal quantities as well as an increased pipe diameter size as an alternative remedy. HYDRA is not an optimization program, thus individual case studies need to be run and analyzed by the engineer.

(page intentionally left blank)

8. ENERGY DISSIPATOR DESIGN

8.1. General Design Concepts

Erosive forces that are at work in the natural drainage network are often increased by construction of a highway. Interception and concentration of overland flow and constriction of natural waterways inevitably results in increased erosion potential. To protect the highway and adjacent areas it is sometimes necessary to employ an energy dissipating device. Energy dissipators should be considered part of the larger design system which includes the culvert and channel protection requirements (upstream and downstream) and may include a debris control structure. The interrelationship of these various components must be considered in designing any one part of the system. For example, energy dissipator requirements may be reduced, increased or possibly eliminated by changes in the culvert design; and downstream channel conditions (velocity, depth and channel stability) will impact the selection and design of appropriate energy dissipation devices.

Throughout the design process, the designer should keep in mind that the primary objective is to protect the highway structure and adjacent area from excessive damage due to erosion. One way to accomplish this objective is to return flow to the downstream channel in a condition that approximates the natural flow regime. Note that this also implies guarding against employing energy dissipation devices that reduce flow conditions substantially below the natural or normal channel conditions. If an energy dissipator is necessary, the first step should be consideration of possible ways of modifying the outlet velocity or erosion potential. This could include modifying the culvert barrel. If an internal modification is not cost effective or is hydraulically unacceptable, the designer must begin the process of selecting and designing an appropriate external energy dissipation device. The following sections summarize some of the factors involved in designing an energy dissipator. For a comprehensive treatment of energy dissipator design.[28]

8.2. Erosion Hazards

Erosion at a culvert inlet is not typically a major problem. At the design discharge, water will normally pond at the inlet, and the only significant increases in velocity will occur upstream of the culvert a distance about equal to the height of the culvert. The average velocity near the inlet may be approximated by dividing the flow rate by the area of the culvert opening. The risk of erosion approaching the inlet should be based on this velocity estimate. Note that the erosion risk may be greater at flow rates less than the design discharge, since depth of ponding at the inlet will be less and greater velocities may occur. This is especially true in channels with steep slopes and high velocity flow.

Most inlet failures have occurred on large flexible-type pipe culverts with projected or mitered entrances without headwalls or other entrance protection. Projecting inlets can bend or buckle from buoyant forces. Mitered entrance edges can be bent in from hydraulic forces. To aid in preventing these types of failures, protective features should include concrete headwalls and/or slope paving.

Erosion at culvert outlets is a common problem. Determination of the flow condition, scour potential and channel erodibility should be standard procedure in the design of all highway culverts. Ultimately, the only safe procedure is to design on the basis that erosion at a culvert outlet and downstream channel will occur, and must be protected against.

8.3. Culvert Outlet Velocity and Velocity Modification

The continuity equation (equation 3) can be used in all situations to compute culvert outlet velocity, either within the barrel or at the outlet. Given the design discharge, the only other information needed is the flow area, and it is a function of the type of control (outlet or inlet).

Culvert outlet velocity is one of the primary indicators of erosion potential. Outlet velocities are seldom less than 3 m/s (10 ft/s) and will range up to 10 m/s (30 ft/s) for culverts on small or mild slopes, and will be even greater for culverts on steep slopes. If the velocity is higher than in the downstream channel, measures to modify or reduce velocity within the culvert barrel should be considered. However, the degree of velocity reduction is typically limited and must be balanced against the increased costs generally involved.

8.3.1. Culverts on Mild Slopes

For culverts on mild slopes operating under outlet control with high tailwater (figures 45a and 45b), the outlet velocity will be determined using the full area of the barrel. With this condition it is possible to reduce the velocity by increasing the culvert size. Note that with high tailwater conditions, erosion may not be a serious problem since the ponded water will act as an energy dissipator; however, it will be important to determine if tailwater will always control, or if any of the other conditions shown on figure 45 might occur.

When the discharge is high enough to produce a critical depth equal to the crown of the culvert barrel (figure 45c), full flow will again occur and the outlet velocity will be based on the area of the barrel. As before, the barrel size can be increased to achieve a reduction in velocity, but it will be necessary to evaluate if the increased size results in a flow depth below the crown, indicating less than full flow at the outlet. When this occurs, the area used in the continuity equation should be based on the actual flow area.

When culverts discharge with critical depth occurring near the outlet (figures 45d and 45e), increasing the barrel size will typically not significantly reduce the outlet velocity. Similarly, increasing the resistance factor will not affect outlet velocity since critical depth is not a function of n.

8.3.2. Culverts on Steep Slopes

For culverts flowing on steep slopes with no tailwater (figures 44a and 44c) the outlet velocity can be determined from normal depth calculations. With normal depth conditions on a steep slope, increasing the barrel size may slightly decrease the outlet velocity; however, calculations show that in reality, the slope is the driving force in establishing the normal depth. The velocity will not be significantly altered by even doubling the culvert size/width. Thus, such an approach is not cost effective. Some reduction in outlet velocity can be obtained by increasing the number of barrels, but this is also generally not cost effective.

Increasing the barrel resistance can significantly reduce outlet velocity and is an important factor in velocity reduction for culverts on steep slopes. The objective is to force full flow near the outlet without creating additional headwater. HEC-14 discusses various methods of creating additional roughness, from changing pipe material to baffles and roughness rings, and details the appropriate design procedures.[28]

8.4. Hydraulic Jump Energy Dissipators

The hydraulic jump is a natural phenomenon which occurs when supercritical flow changes to subcritical flow (see chapter 4). This abrupt change in flow condition is accomplished by considerable turbulence and loss of energy, making the hydraulic jump an effective energy dissipation device. To better define the location and length of a hydraulic jump, standard design structures have been developed to force the hydraulic jump to occur. These structures typically use blocks, sills or other roughness elements to impose exaggerated resistance to flow. Forced hydraulic jump structures applicable in highway engineering include the Colorado State University (CSU) rigid boundary basin, USBR type IV basin and the St. Anthony Falls basin.[28]

The CSU rigid boundary basin was developed from model study tests of basins with abrupt expansions (figure 49); however, the configuration recommended for use is a combination flared-abrupt expansion basin. The roughness elements are symmetrical about the basin centerline and the spacing between the elements is approximately equal to the element width. Alternate rows of roughness elements are staggered. Riprap may be needed for a short distance downstream of the basin.

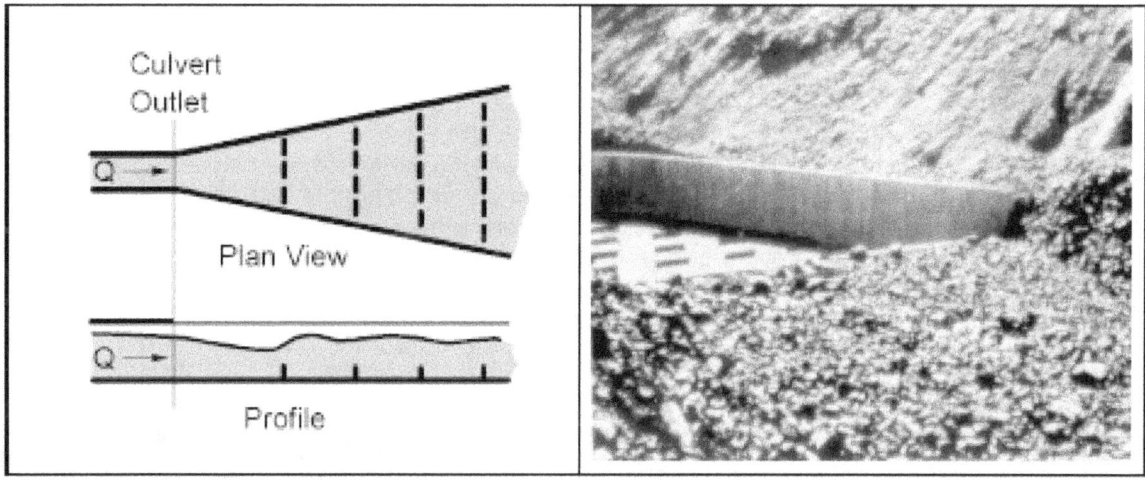

Figure 49a. Schematic of CSU rigid   Figure 49b. CSU rigid boundary basin.
            boundary basin.

The St. Anthony Falls (SAF) stilling basin is a more generalized design that uses special appurtenances, chute blocks and baffle or floor blocks to force the hydraulic jump to occur (figure 50). It is recommended for Froude Numbers between 1.7 and 17. Similar to the CSU basin, the design criteria were developed from model study test results.

8.5. Impact Basins

As the name implies, impact basins are designed with part of the structure physically blocking the free discharge of water. The action of water impacting on the structure dissipates energy and modifies the downstream flow regime. Impact basins include the Contra Costa Energy Dissipator, Hook type energy dissipator, and the USBR Type VI Stilling Basin.[28]

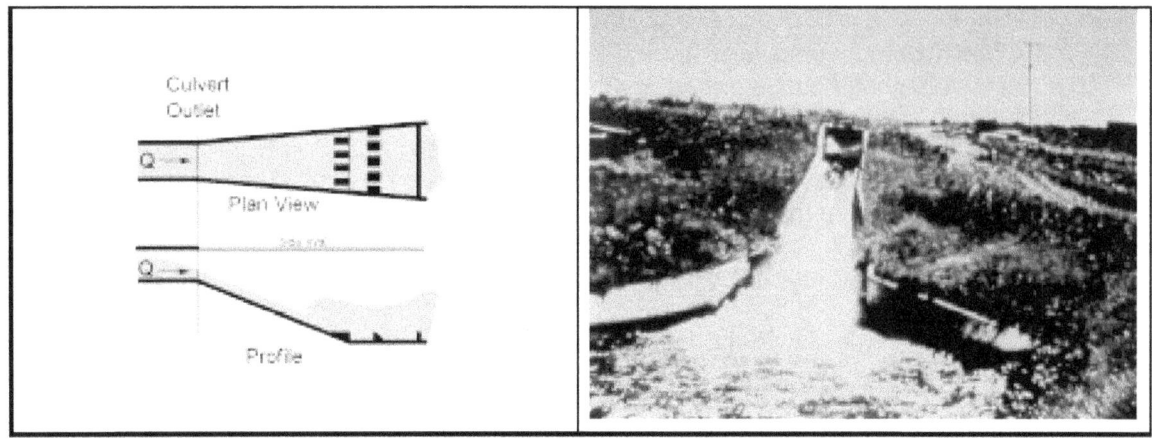

Figure 50a. Schematic of SAF
stilling basin.

Figure 50b. SAF stilling basin.

The impact basin most commonly used in highway engineering is probably the USBR Type VI (figure 51). The structure is contained in a relatively small box-like structure which requires no tailwater for successful performance. The shape of the basin evolved from extensive tests, and resulted in a design based around a vertical hanging baffle. Energy dissipation is initiated by flow striking the vertical hanging baffle and being deflected upstream by the horizontal portion of the baffle and by the floor, creating horizontal eddies. Notches in the baffle provide a self cleaning feature after prolonged nonuse of the structure. If the basin is full of sediment, the notches provide concentrated jets of water for cleaning, and if the basin is completely clogged the full discharge can be carried over the top of the baffle. Use of the basin is limited to installations where the velocity at the entrance of the basin does not exceed 15 m/s (50 ft/s) and discharge is less than 11 m^3/s (400 ft^3/s).

Figure 51a. Schematic of USBR
Type IV.

Figure 51b. Baffle-wall energy dissipator -
USBR Type VI.

158

8.6.   Drop Structures with Energy Dissipation

Drop structures are commonly used for flow control and energy dissipation. Reducing channel slope by placing drop structures at intervals along the channel changes a continuous steeper sloped channel into a series of milder sloped reaches with vertical drops. Instead of slowing down and transferring high erosion producing velocities into lower nonerosive velocities, drop structures control the slope of the channel so that high velocities never develop. The kinetic energy or velocity gained by the water as it drops over the crest of each structure is dissipated by specially designed aprons or stilling basins.

Energy dissipation occurs through impact of the falling water on the floor, redirection of the flow, and turbulence. The stilling basin used to dissipate excess energy can vary from a simple concrete apron to an apron with flow obstructions such as baffle blocks, sills, or abrupt rises.   The length of the concrete apron required can be shortened by addition of these appurtenances. Figure 52 illustrates a straight drop stilling basin with floor blocks and an end sill. The design of this and other drop structure stilling basins is detailed in HEC-14.[28]

8.7.   Stilling Wells

Stilling wells dissipate kinetic energy by forcing flow to travel vertically upward to reach the downstream channel. The stilling well most commonly used in highway engineering is the Corps of Engineers Stilling Well (figure 53). This stilling well has application where debris is not a serious problem. It will operate with moderate to high concentrations of sand and silt, but is not recommended for areas where quantities of large floating or rolling debris are expected unless suitable debris-control structures are used. Its greatest application in highway engineering is at the outfalls of storm drains and pipe down drains where little debris is expected. It is recommended that riprap or other types of channel protection be provided around the stilling well outlet.

8.8.   Riprap Stilling Basins

Riprap stilling basins are commonly used at culvert outfalls (figure 54). The design procedure for riprap energy dissipators was developed from model study tests. The results of this testing indicated that the size of the scour hole at the outlet of a culvert was related to the size of the riprap, discharge, brink depth and tailwater depth. The mound of rock material that often forms on the bed downstream of the scour hole contributes to dissipation of energy and reduces the size of scour hole. The general design guidelines for riprap stilling basins include preshaping the scour hole and lining it with riprap. Specific design criteria for the length, depth and width of the scour hole, and the entire basin, are provided in HEC-14.[28]

8.9.   Energy Dissipator Design Using HY-8

Energy dissipator design for culvert outlets can be completed within the HY-8 module of HYDRAIN.[10] The design is based on FHWA publication HEC-14.[28] A performance curve is necessary to perform the energy dissipator design and analysis.

159

Figure 52. Straight drop spillway stilling basin.

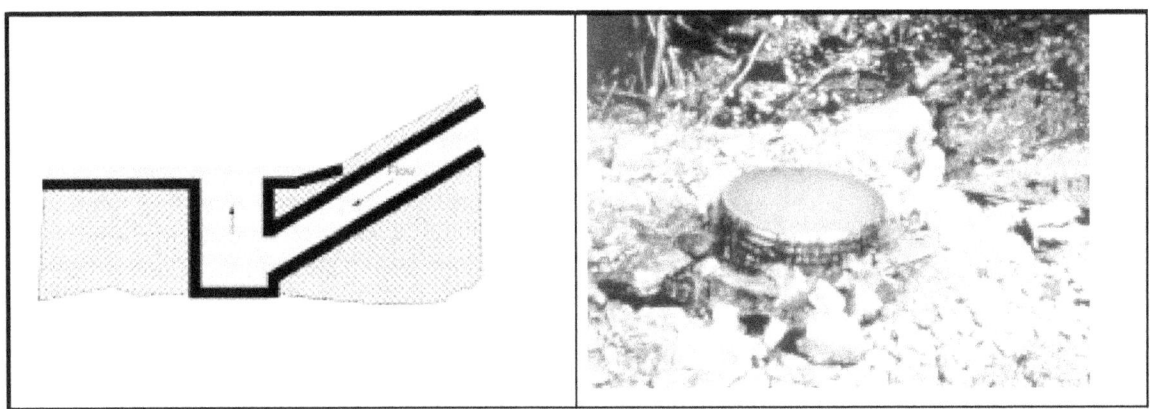

Figure 53a. Schematic of COE stilling well.

Figure 53b. COE stilling well.

Figure 54. Riprapped culvert energy basin.

9. DRAINAGE SYSTEM CONSTRUCTION

9.1. General

Construction methods depend upon the equipment used and the expertise of the contractor and are outside the scope of this publication. This chapter will discuss a few of the procedures that should be followed to construct satisfactory highway drainage facilities.

Drainage facilities should be constructed early in the grading operations and any necessary erosion protection should be provided before potential damage occurs. Effective drainage during construction frequently eliminates costly delays as well as later failures that might result from a saturated subgrade. Slopes should be protected from erosion as early as practicable in order to minimize damage and lessen the discharge of eroded soil into existing and newly constructed drainage facilities.

9.2. Supervision

Proper design of drainage facilities will not produce an adequate drainage system without careful supervision during construction. Supervision of drainage structure construction means not only seeing that the construction complies with the plans and specifications, but that any omissions in the plans are corrected. Drainage facilities should be shown on the construction plans together with sufficient hydraulic design data, such as drainage area and design discharge, so that the necessary information is available to solve unforeseen future drainage problems.

9.3. Excavation

Drainage facilities are usually placed from the outlet toward the higher end so that the channel will drain during construction. Dikes for intercepting channels are preferably built from material excavated from the adjacent cuts without disturbing the natural soil at the channel location.

9.4. Grass-lined Channels

Grass lining can best be attained by sodding. The upper parts of the channel may be sprigged or seeded if the cost of sod makes this necessary, but the time of the year and the likelihood of damaging rains occurring before the seedlings become established, should be considered. A type of grass should be selected that is adapted to the locality and to the site conditions. Where sod grasses are viable, they are preferred to bunch grasses because of their superior performance in resisting erosion.

Seeding can be protected by mulch, temporary cover grasses, geotextiles, etc. Sod strips perpendicular to the channel centerline at regular intervals have also been used to protect the intervening seeded area. Sod might also be used in the channel bottom and up part of the sideslope for immediate protection with the remainder of channel slope seeded.

161

9.5. Concrete-lined Channels and Chutes

The use of rigid linings is discouraged; however, if used, concrete channel linings can be cast-in-place, shotcrete, or precast. Soil-cement linings have been successful at some locations. Concrete linings must be placed on a firm well-drained foundation to prevent cracking or failure of the lining. Soil of low density should be thoroughly compacted or removed and replaced with suitable material. Where the soil is deep loess, concrete or other type rigid linings might not be suitable. Expansive clays are extremely hazardous to rigid-type linings because their movement buckles the linings as well as producing an unstable support.

When placing an unformed slab on a slope, a tendency exists to use a stiff concrete mix that will not slough; however, experience indicates that placement of such low-slump concrete without thorough vibration usually results in considerable honeycombing on the underside.[13] To avoid such results, the concrete should not be stiffer than a 63 mm slump. Concrete of this consistency will barely stay on a steep slope. After spreading, the concrete should be thoroughly vibrated, preferably just ahead of a weighted steel-faced slipform screen working up the slope.[13]

The linings of channels that carry high-velocity flow should be poured as nearly monolithic as possible, without expansion joints or weepholes, and using as few construction joints as possible. Construction joints should be made watertight. Longitudinal and transverse reinforcing steel should be used throughout to control cracking with the longitudinal steel carried through the construction joints. The lining should be anchored to the slope as necessary by reinforced cutoff walls to prevent sliding.

Proper curing of the concrete lining is important, particularly in warm, dry, windy weather, to prevent the early drying of corners, edges, and surfaces. A well-moistened subgrade and wet burlap in contact with the exposed concrete surfaces is excellent for curing purposes.

The edges of newly constructed channels should be protected by a strip of sod at the time of construction. The design of the lining edges should allow enough depth of soil to permit the growth of grass.

9.6. Bituminous-lined Channels

A well-drained subgrade is necessary beneath a bituminous lining because the strength and weight of the lining is not sufficient to withstand high hydrostatic uplift pressure. Weed control measures are sometimes necessary before the lining is placed. These measures consist of careful grubbing of the subgrade followed by the application of soil sterilant.

9.7. Riprap-lined Channels

All stone used for channel linings or bank protection should be hard, dense, and durable. Most of the igneous and metamorphic rocks, many of the limestones, and some of the sandstones make excellent linings. Shale is not suitable, and limestones and sandstones that have shale seams are undesirable. Quarried stones, angular in shape, are preferred to rounder boulders or cobbles. HEC-15 provides guidelines for riprap gradation, thickness and filter requirements for roadside channels and HEC-11 for bank protection.[26,29]

The stones should be placed on the filter blanket or prepared natural slope in a manner which will produce a reasonably well-graded mass of stone with the minimum practicable percentage of voids. Stone protection should be placed to its full course thickness at one operation and in such a manner as to avoid displacing the underlying material. Placing of stone protection in layers or by dumping into chutes or by similar methods likely to cause segregation should not be permitted. The larger stones should be well distributed and the entire mass of stones should roughly conform to the gradation specified. The stone protection should be so placed and distributed as to avoid large accumulations or areas composed largely of either the larger or smaller sizes of stone. The mass should be fairly compact, with all sizes of material placed in their proper proportions. Hand-placing and rearranging of individual stones by mechanical equipment may be required to the extent necessary to secure the results specified above. The desired distribution of the various sizes of stone throughout the mass might be obtained by selective loads during placing, or by a combination of these methods. Ordinarily, the stone protection should be placed in conjunction with the construction of the embankment with only sufficient delay in construction of the stone protection as may be necessary to prevent mixture of embankment and stone.

Bank protection where the channel is composed of sand or silt, should extend a minimum vertical distance of 1.5 m (5 ft) below the streambed on a continuous slope with the embankment. Where the streambed is of material other than sand or silt, the bank protection should be terminated in a rock toe at the level of the streambed to prevent undermining the bank protection. The toe should have a minimum base width of 1.5 m (5 ft) and a minimum thickness of 1 m (3 ft) with 1 and 1-1/2 side slopes. On large rivers, or tidal estuaries, having a considerable depth of flow at low water stages, the stone protection should extend a short distance (e.g., 1.5 m (5 ft)) below the mean low water and the toe can be omitted.

Hand-placed stone should be carefully laid to produce a more or less definite pattern with a minimum of voids and with the top surface relatively smooth. Joints should be staggered between courses. The stone used for hand-placed protection should be of better quality than the minimum quality suitable for dumped stone protection. Stones that are roughly square and of fairly uniform thickness are much easier to place than irregular stones. Stone of a flat stratified nature should be placed with the principal bedding planes normal to the slope. Openings to the subsurface should be filled with rock fragments; however, enough voids or openings should be left to drain the subsurface properly.

(page intentionally left blank)

10. DRAINAGE SYSTEM MAINTENANCE

10.1. General

Drainage facilities rapidly lose their effectiveness unless they are adequately maintained. Thus, a good maintenance program is of equal importance with the proper design and construction of the drainage system. In fact, a knowledge of the equipment to be used in maintenance and the methods to be employed is a prerequisite to proper design.

Maintenance of vegetative cover on slopes and in drainage channels requires continued attention. The original treatment applied during construction will not last forever. Repeated applications of fertilizer, lime, or organic material at intervals are as necessary on the highway roadside as on the home lawn. Areas often need reseeding or resodding to restore the vegetative cover. This should be done before serious erosion occurs.

Minor erosion damage within the highway right-of-way should be repaired immediately after it occurs and action taken to prevent a recurrence of the damage. The damage caused by light storms reveals the points of weakness in the drainage system. If these weaknesses are corrected when repairing the damage itself, the drainage system will likely carry the design discharge without damage. Deficiencies that are found in the drainage systems and the corrective action taken should be reported to the hydraulic or design engineer so that similar troubles will not occur on future construction. Reports on drainage works that function well during severe storms are equally valuable to the designer.

10.2. Effect of Maintenance on Flow Capacity

Maintenance of highway drainage facilities includes repairing erosion damage, mowing grass-lined channels, and removing any deposited sediment or debris. All these measures keep the capacity of the drainage system at the design level. If a channel or culvert contains brush, sediment, or debris, the flow capacity will be less than the design value. In a grass-lined channel, deposited sediment and debris may kill the vegetative lining with subsequent erosion damage during higher flood flows. In some situations, sediment traps and debris barriers might be constructed in order to collect the objectionable material for easy removal.

The effect of inadequate maintenance in a grass-lined channel can be illustrated by considering the Manning's equation and the effect of vegetation on n values. For a grass-lined channel that is regularly mowed, the n value would be relatively low, e.g., 0.035. In contrast, if the same channel is not mowed and grass/weeds are allowed to grow, the n value would be relatively high, e.g., 0.10. As a result, the channel will only carry 0.035/0.10 = 0.35, or about one-third the flow for which it was designed. The remainder of the design flow would overflow the channel and cause flooding or possible erosion.

(page intentionally left blank)

11.   DRAINAGE SYSTEM ECONOMICS

11.1.   General

Providing adequate drainage is essential to the existence of the highway. Economical drainage design is achieved through doing an adequate job at the lowest cost.

The lowest cost adequate drainage system maintains proper balance between first cost, flood damage, and maintenance cost, and has the capacity and protection to carry the runoff for which it was designed. Selection of the frequency of the design runoff is a matter of economics, while estimating the magnitude of the storm runoff for a selected frequency belongs in the field of hydrology. In this chapter, some of the factors to be considered in the economic selection of highway drainage facilities are discussed.

11.2.   Frequency of the Design Storm

The average annual cost of a drainage facility is the sum of (1) the first cost divided by the expected life of the drainage facility, plus (2) the average annual maintenance cost, plus (3) the annual charge for possible damage from facility runoff exceeding the design capacity. The average annual maintenance cost might also include the annual charge for flood damage if a flood exceeding the design capacity has occurred; however, it is probably better to separate these costs. The damage to a facility designed to carry a 10-year runoff from a chance occurrence of, say, a 200-year runoff in a few years' record of maintenance expenditures would distort the average annual maintenance cost. The annual charge for possible flood damage should consider the frequency of the design storm and equals the cost of damage from runoff exceeding the design capacity divided by the return period, in years, of the design storm.

If these costs could be evaluated for various combinations of the component costs, the most economical drainage system could be determined as the one with the lowest average annual cost. The optimum frequency of the design storm would then be the frequency associated with the storm runoff that, in combination with other costs, produced the lowest average annual cost. However, the three cost items are interrelated and difficult to evaluate, particularly the item of damage by runoff that exceeds the design runoff. The cost of storm damage includes the cost of traffic interruption by floodwaters or washed-out highway, as well as the cost of repairing the damage to the highway and drainage system and the additional damage to the abutting property directly attributable to the presence of the highway. A further complication is the variation in damage due to the magnitude by which the runoff exceeds the design runoff.

Individual analysis for each small drainage system is impractical, if not impossible. The best solution appears to be a study of average conditions and selection of the frequency of design runoff to be used for various drainage structures according to the class of the highway. The designated frequencies might vary from state to state or even within a state composed of areas differing widely in topography or population density. Individual variations in the designated frequency of the design storm might be needed at locations where damage by flooding could be great, but the cost of a larger facility to carry a less frequent storm is moderate. Economic analysis as applied to drainage structure design is discussed in HEC-17.[30]

(page intentionally left blank)

12. REFERENCES

1. McCuen, R.H., Johnson, P.A., and Ragan, R.M., 1996, "Highway Hydrology," U.S. Department of Transportation, Federal Highway Administration, Hydraulic Design Series No. 2, FHWA-SA-96-067.

2. Interagency Advisory Committee on Water Data, 1982, "Flood Flow Frequency," Bulletin 17B of the Hydrology Subcommittee.

3. U.S. Weather Bureau, 1961, "Rainfall-Frequency Atlas of the United States," Tech. Paper 40, Washington, D.C.

4. AASHTO, 1992, Hydraulic Drainage Guidelines, Edition 2, Volumes 1 through 10 and Glossary.

5. ASCE, 1992, "Design and Construction of Urban Stormwater Management Systems."

6. Brown, S.A., Stein, S.M., and Warner, J.C., 1996, "Urban Drainage Design Manual," U.S. Department of Transportation, Federal Highway Administration, Hydraulic Engineering Circular No. 22, FHWA-SA-96-078.

7. Soil Conservation Service, 1986, "Urban Hydrology for Small Watersheds," U.S. Department of Agriculture, Engineering Division, Technical Report No. 55.

8. Sauer V.B., Thomas, W.D., Jr., Stricker, V.A., and Wilson, K.V., 1983, "Flood Characteristics of Urban Watersheds in the United States," U.S. Geological Survey, Water Supply Report No. 2207, 63 pp.

9. U.S. Geological Survey, 1994, Nationwide Summary U.S. Geological Survey Regional Regression Equations for Estimating Magnitude and Frequency of Floods for Ungaged Site, 1993. Water Resources Investigations Report 94-4002, compiled by M.E. Jenning, W.O. Thomas, and H.C. Riggs, Water Resources Investigations Report No. 94-4002.

10. Federal Highway Administration, 1994, "HYDRAIN, Integrated Drainage Design Computer System: Version 5.0," Volumes I to VII, Report No. FHWA-RD-92-061 (software is available from McTRANS Center, Gainesville, Florida).

11. Richardson, E.V., Simons, D.B., and Julien, P., 1990, "Highways in the River Environment," Report No. FHWA-HI-90-016, NTIS PB90-252479, Federal Highway Administration, U.S. Department of Transportation, Washington, D.C.

12. Brater, E.F. and King, H.W., 1976, "Handbook of Hydraulics," McGraw-Hill Book Company, Inc., New York, NY, 6th ed.

13. U.S. Bureau of Reclamation, 1973, "Design of Small Dams," U.S. Government Printing Office, Washington, D.C., 816 pp.

14. Bradley, J.N., Revised 1978, "Hydraulics of Bridge Waterways," Hydraulic Design Series No. 1, Federal Highway Administration, U.S. Government Printing Office, Washington, D.C., FHWA-EPD86-101, NTIS PB86-181708, 2nd edition 1970, 111 pp.

15. Richardson, E.V. and Davis, S.R., 2001. "Evaluating Scour at Bridges," Report No. FHWA-NHI-01-001, Hydraulic Engineering Circular No. 18, U.S. Department of Transportation, Federal Highway Administration, Washington, D.C.

16. Lagasse, P.F., Schall, J.D., and Richardson, E.V., 2001, "Stream Stability at Highway Structures," U.S. Department of Transportation, Report No. FHWA-NHI-01-002, Hydraulic Engineering Circular No. 20, Federal Highway Administration, Washington, D.C.

17. Chow, V.T., 1959, "Open Channel Hydraulics," McGraw-Hill Book Company, Inc., New York, NY, 680 pp.

18. Jarrett, R.D., 1985, "Determination of Roughness Coefficients for Streams in Colorado," U.S. Geological Survey, Water Resources Investigations Report No. 85-4004.

19. Cowan, W.L., 1956, "Estimating Hydraulic Roughness Coefficients, Agricultural Engineering, v. 37, no. 7, pp. 473-475.

20. Benson, M.A. and Dalrymple, T., 1967, "General Field and Office Procedures for Indirect Discharge Measurements," USGS Techniques of Water Resources Investigations, Book 3, Chap. A1, 30 p.

21. Aldridge, B.N. and Garrett, J.M., 1973, "Roughness Coefficients for Streams in Arizona," USGS Open File Report, 87 p.

22. Arcement, G.K. and V.R. Schneider, 1984, "Guide for Selecting Manning's Roughness Coefficients for Natural Channels and Floodplains," USGS Water Supply Paper 2339.

23. Einstein, H. A., 1950. "The Bed-Load Function for Sediment Transportation in Open-Channel Flows." U.S. Department of Agriculture, Soil Conservation Service, Technical Bulletin No. 1026.

24. Rouse, H., ed., 1950,"Engineering Hydraulics," John Wiley & Sons, Inc., New York, NY, 1039 pp.

25. AASHTO, "Roadside Design Guide," 1996.

26. Chen, Y.H. and Cotton, G.K., 1988. "Design of Roadside Channels with Flexible Linings," U.S. Department of Transportation, Federal Highway Administration, FHWA IP87-7, NTIS PB89-122584, Hydraulic Engineering Circular No. 15.

27. Normann, J.M., Houghtalen, R.J., and Johnston, W.J., 1985, "Hydraulic Design of Highway Culverts, U.S. Department of Transportation, Federal Highway Administration, FHWA IP-85-15, NTIS PB86-196961, Hydraulic Design Series No. 5.

28. Corry, M.L., Thompson, P.L., Watts, F.J., Jones, J.S., and Richards, D.L., 1983. "The Hydraulic Design of Energy Dissipators for Culverts and Channels," U.S. Department of Transportation, Federal Highway Administration, FHWA EPD-86-110, NTIS PB86-180205, Hydraulic Engineering Circular No. 14.

29. Federal Highway Administration, 1989, "Design of Riprap Revetment," Report No. FHWA-IP-89-016, NTIS PB89-218424, Hydraulic Engineering Circular No. 11, Federal Highway Administration, U.S. Department of Transportation, Washington, D.C.

30.   Federal Highway Administration, 1981, "The Design of Encroachments of Flood Plains Using Risk Analysis," Report No. FHWA-EPD-86-1112 NTIS PB86-182110, Hydraulic Engineering Circular No. 17.

ADDITIONAL REFERENCES

1.   American Association of State Highway Officials, "A Policy on Geometric Design of Highways," Washington, D.C., 1994, 1,050 pp.

2.   Federal Highway Administration, "Highway Subdrainage Design," FHWA-TS-80-224, Government Printing Office 050-001-00195-1, 1980, 162 pp.

3.   U.S. Soil Conservation Service, Hydrology, Section 4, including Supplement A, Soil Conservation Service National Engineering Handbook, Washington, D.C.

4.   U.S. Bureau of Reclamation, "Hydraulic Design of Stilling Basins and Energy Dissipators," Engineering Monograph No. 25, U.S. Government Printing Office, Washington, D.C., 1964, 224 pp.

5.   Masch, F.D., 1984, "Hydrology," U.S. Department of Transportation, Federal Highway Administration, FHWA IP 84-15, NTIS PB85-182954, Hydraulic Engineering Circular 19, converted to Hydraulic Design Series 2, "Highway Hydrology," 1996.

6.   Federal Highway Administration, 1971, "Debris-Control Structures," Report No. FHWA-EPD-86-106, NTIS PB86-179801, Hydrologic Engineering Circular No. 9.

7.   Federal Highway Administration, 1993, "Bridge Deck Drainage Systems," Report No. FHWA-SA-92-101, NTIS PB94-109584, Hydraulic Engineering Circular No. 21.

8.   Federal Highway Administration, 1984, "Guide for Selecting Manning's Roughness Coefficient for Natural Channels and Flood Plains," Report No. FHWA-TS-84-204, NTIS-PB84-242585.

9.   AASHTO, 1991, Model Drainage Manual, First Edition (U.S Customary Units) and 1999 Metric Edition, available from AASHTO (202-624-5800).

10.   AASHTO, 1999, "Highway Drainage Guidelines," Metric Edition, Volumes 1 through 14 and Glossary, available at AASHTO (202-624-5800).

11.   Barnes, H.H., 1967, "Roughness Characteristics of Natural Channels," USGS Water Supply Paper 1849.

(page intentionally left blank)

APPENDIX A

USE OF THE METRIC SYSTEM

(page intentionally left blank)

APPENDIX A

USE OF THE METRIC SYSTEM

The following information is summarized from the Federal Highway Administration, National Highway Institute (NHI) Course No. 12301, "Metric (SI) Training for Highway Agencies." For additional information, refer to the Participant Notebook for NHI Course No. 12301.

In SI there are seven base units, many derived units and two supplemental units (table 3). Base units uniquely describe a property requiring measurement. One of the most common units in civil engineering is length, with a base unit of meters in SI. Decimal multiples of meter include the kilometer (1000 m), the centimeter (1 m/100) and the millimeter (1 m/1000). The second base unit relevant to highway applications is the kilogram, a measure of mass which is the inertial of an object. There is a subtle difference between mass and weight. In SI, mass is a base unit, while weight is a derived quantity related to mass and the acceleration of gravity, sometimes referred to as the force of gravity. In SI the unit of mass is the kilogram and the unit of weight/force is the newton. Table 4 illustrates the relationship of mass and weight. The unit of time is the same in SI as in the English system (seconds). The measurement of temperature is Centigrade. The following equation converts Fahrenheit temperatures to Centigrade, $°C = 5/9 (°F - 32°)$.

Derived units are formed by combining base units to express other characteristics. Common derived units in highway drainage engineering include area, volume, velocity, and density. Some derived units have special names (table 5).

Table 6 provides useful conversion factors from English to SI units. The symbols used in this table for metric units, including the use of upper and lower case (e.g., kilometer is "km" and a newton is "N") are the standards that should be followed. Table 7 provides the standard SI prefixes and their definitions.

Table 8 provides physical properties of water at atmospheric pressure in SI system of units. Table 9 gives the sediment grade scale and table 10 gives some common equivalent hydraulic units.

Table 3. Overview of SI Units.		
	Units	Symbol
Base units		
length	meter	m
mass	kilogram	kg
time	second	s
temperature*	kelvin	K
electrical current	ampere	A
luminous intensity	candela	cd
amount of material	mole	mol
Derived units		
Supplementary units		
angles in the plane	radian	rad
solid angles	steradian	sr
*Use degrees Celsius (°C), which has a more common usage than kelvin.		

Table 4. Relationship of Mass and Weight.			
	Mass	Weight or Force of Gravity	Force
English	slug pound-mass	pound pound-force	pound pound-force
metric	kilogram	newton	newton

Table 5. Derived Units With Special Names.			
Quantity	Name	Symbol	Expression
Frequency	hertz	Hz	s^{-1}
Force	newton	N	$kg \cdot m/s^2$
Pressure, stress	pascal	Pa	N/m^2
Energy, work, quantity of heat	joule	J	$N \cdot m$
Power, radiant flux	watt	W	J/s
Electric charge, quantity	coulomb	C	$A \cdot s$
Electric potential	volt	V	W/A
Capacitance	farad	F	C/V
Electric resistance	ohm	Ω	V/A
Electric conductance	siemens	S	A/V
Magnetic flux	weber	Wb	$V \cdot s$
Magnetic flux density	tesla	T	Wb/m^2
Inductance	henry	H	Wb/A
Luminous flux	lumen	lm	$cd \cdot sr$
Illuminance	lux	lx	lm/m^2

Table 6. Useful Conversion Factors.			
Quantity	From English Units	To Metric Units	Multiplied by*
Length	mile	km	1.609
	yard	m	0.9144
	foot	m	0.3048
	inch	mm	25.40
Area	square mile	km²	2.590
	acre	m²	4047
	acre	hectare	0.4047
	square yard	m²	0.8361
	square foot	m²	0.092 90
	square inch	mm²	645.2
Volume	acre foot	m³	1 233
	cubic yard	m³	0.7646
	cubic foot	m³	0.028 32
	cubic foot	L (1000 cm³)	28.32
	100 board feet	m³	0.2360
	gallon	L (1000 cm³)	3.785
	cubic inch	cm³	16.39
Mass	lb	kg	0.4536
	kip (1000 lb)	metric ton (1000 kg)	0.4536
Mass/unit length	plf	kg/m	1.488
Mass/unit area	psf	kg/m²	4.882
Mass density	pcf	kg/m³	16.02
Force	lb	N	4.448
	kip	kN	4.448
Force/unit length	plf	N/m	14.59
	klf	kN/m	14.59
Pressure, stress, modulus of elasticity	psf	Pa	47.88
	ksf	kPa	47.88
	psi	kPa	6.895
	ksi	MPa	6.895
Bending moment, torque, moment of force	ft-lb	N•m	1.356
	ft-kip	KN•m	1.356
Moment of mass	lb•ft	kg•m	0.1383
Moment of inertia	lb•ft	kg•m	0.042 14
Second moment of area	in⁴	mm⁴	416 200
Section modulus	in³	mm³	16 390
Power	ton (refrig)	kW	3.517
	Btu/s	kW	1.054
	hp (electric)	W	745.7
	Btu/h	W	0.2931

Table 6. Useful Conversion Factors (continued).			
Quantity	From English Units	To Metric Units	Multiplied by*
Volume rate of flow	ft^3/s cfm cfm mgd	m^3/s m^3/s L/s m^3/s	0.028 32 0.000 471 9 0.4719 0.0438
Velocity, speed	ft/s	m/s	<u>0.3048</u>
Acceleration	ft/s^2	m/s^2	<u>0.3048</u>
Momentum	lb•ft/sec	kg•m/s	0.1383
Angular momentum	lb•ft^2/s	kg•m^2/s	0.042 14
Plane angle	degree	rad mrad	0.017 45 17.45
* 4 significant figures; underline denotes exact conversion			

Table 7. Prefixes.					
Submultiples			Multiples		
deci	10^{-1}	d	deka	10^1	da
centi	10^{-2}	c	hecto	10^2	h
milli	10^{-3}	m	kilo	10^3	k
micro	10^{-6}	µ	mega	10^6	M
nano	10^{-9}	n	giga	10^9	G
pica	10^{-12}	p	tera	10^{12}	T
femto	10^{-15}	f	peta	10^{15}	P
atto	10^{-18}	a	exa	10^{18}	E
zepto	10^{-21}	z	zetta	10^{21}	Z
yocto	10^{-24}	y	yotto	10^{24}	Y

Table 8. Physical Properties of Water at Atmospheric Pressure in SI Units.								
Temperature		Density	Specific Weight	Dynamic Viscosity	Kinematic Viscosity	Vapor Pressure	Surface Tension[1]	Bulk Modulus
Celcius	Fahrenheit	kg/m^3	N/m^3	$N \cdot s/m^2$	m^2/s	N/m^2 abs.	N/m	GN/m^2
0°	32°	1,000	9,810	1.79×10^{-3}	1.79×10^{-6}	611	0.0756	1.99
5°	41°	1,000	9,810	1.51×10^{-3}	1.51×10^{-6}	872	0.0749	2.05
10°	50°	1,000	9,810	1.31×10^{-3}	1.31×10^{-6}	1,230	0.0742	2.11
15°	59°	999	9,800	1.14×10^{-3}	1.14×10^{-6}	1,700	0.0735	2.16
20°	68°	998	9,790	1.00×10^{-3}	1.00×10^{-6}	2,340	0.0728	2.20
25°	77°	997	9,781	8.91×10^{-4}	8.94×10^{-7}	3,170	0.0720	2.23
30°	86°	996	9,771	7.97×10^{-4}	8.00×10^{-7}	4,250	0.0712	2.25
35°	95°	994	9,751	7.20×10^{-4}	7.24×10^{-7}	5,630	0.0704	2.27
40°	104°	992	9,732	6.53×10^{-4}	6.58×10^{-7}	7,380	0.0696	2.28
50°	122°	988	9,693	5.47×10^{-4}	5.53×10^{-7}	12,300	0.0679	
60°	140°	983	9,643	4.66×10^{-4}	4.74×10^{-7}	20,000	0.0662	
70°	158°	978	9,594	4.04×10^{-4}	4.13×10^{-7}	31,200	0.0644	
80°	176°	972	9,535	3.54×10^{-4}	3.64×10^{-7}	47,400	0.0626	
90°	194°	965	9,467	3.15×10^{-4}	3.26×10^{-7}	70,100	0.0607	
100°	212°	958	9,398	2.82×10^{-4}	2.94×10^{-7}	101,300	0.0589	
[1]Surface tension of water in contact with air								

Table 9. Sediment Particles Grade Scale.						
Size				Approximate Sieve Mesh Openings Per Inch		Class
Millimeters		Microns	Inches	Tyler	U.S. Standard	
4000-2000	----	----	160-80	----	----	Very large boulders
2000-1000	----	----	80-40	----	----	Large boulders
1000-500	----	----	40-20	----	----	Medium boulders
500-250	----	----	20-10	----	----	Small boulders
250-130	----	----	10-5	----	----	Large cobbles
130-64	----	----	5-2.5	----	----	Small cobbles
64-32	----	----	2.5-1.3	----	----	Very coarse gravel
32-16	----	----	1.3-0.6	----	----	Coarse gravel
16-8	----	----	0.6-0.3	2 1/2	----	Medium gravel
8-4	----	----	0.3-0.16	5	5	Fine gravel
4-2	----	----	0.16-0.08	9	10	Very fine gravel
2-1	2.00-1.00	2000-1000	----	16	18	Very coarse sand
1-1/2	1.00-0.50	1000-500	----	32	35	Coarse sand
1/2-1/4	0.50-0.25	500-250	----	60	60	Medium sand
1/4-1/8	0.25-0.125	250-125	----	115	120	Fine sand
1/8-1/16	0.125-0.062	125-62	----	250	230	Very fine sand
1/16-1/32	0.062-0.031	62-31	----	----	----	Coarse silt
1/32-1/64	0.031-0.016	31-16	----	----	----	Medium silt
1/64-1/128	0.016-0.008	16-8	----	----	----	Fine silt
1/128-1/256	0.008-0.004	8-4	----	----	----	Very fine silt
1/256-1/512	0.004-0.0020	4-2	----	----	----	Coarse clay
1/512-1/1024	0.0020-0.0010	2-1	----	----	----	Medium clay
1/1024-1/2048	0.0010-0.0005	1-0.5	----	----	----	Fine clay
1/2048-1/4096	0.0005-0.0002	0.5-0.24	----	----	----	Very fine clay

Table 10. Common Equivalent Hydraulic Units.								
Volume								
Unit	Equivalent							
	Cubic Inch	Liter	U.S. Gallon	Cubic Foot	Cubic Yard	Cubic Meter	Acre-Foot	Sec-Foot-Day
Liter	61.02	1	0.264 2	0.035 31	0.001 308	0.001	810.6 E - 9	408.7 E - 9
U.S. Gallon	231.0	3.785	1	0.133 7	0.004 951	0.003 785	3.068 E - 6	1.547 E - 6
Cubic Foot	1,728	28.32	7.481	1	0.037 04	0.028 32	22.96 E - 6	11.57 E - 6
Cubic Yard	46,660	764.5	202.0	27	1	0.746 6	619.8 E - 6	312.5 E - 6
Meter^3	61,020	1,000	264.2	35.31	1.308	1	810.6 E - 6	408.7 E - 6
Acre-Foot	75.27 E + 6	1,233,000	325,900	43 560	1 613	1 233	1	0.504 2
Sec-Foot-Day	149.3 E + 6	2,447,000	646,400	86 400	3 200	2 447	1,983	1
Discharge (Flow Rate, Volume/Time)								
Unit		Gallon/Min	Liter/Sec	Acre-Foot/Day	Foot^3/Sec	Million Gal/Day	Meter^3/Sec	
Gallon/Minute		1	0.063 09	0.004 419	0.002 228	0.001 440	63.09 E - 6	
Liter/Second		15.85	1	0.070 05	0.035 31	0.022 82	0.001	
Acre-Foot/Day		226.3	14.28	1	0.504 2	0.325 9	0.014 28	
Feet^3/Second		448.8	28.32	1.983	1	0.646 3	0.028 32	
Meter^3/Second		15,850	1,000	70.04	35.31	22.82	1	

APPENDIX B

Drainage Design Charts and Tables

(page intentionally left blank)

APPENDIX B

Drainage Design Charts and Tables

Table 11. Values of Runoff Coefficients, C, for Use in the Rational Method.	
Type of Surface	Runoff Coefficient (C[1])
Rural Areas	
Concrete or sheet asphalt pavement	0.8 - 0.9
Asphalt macadam pavement	0.6 - 0.8
Gravel roadways or shoulders	0.4 - 0.6
Bare earth	0.2 - 0.9
Steep grassed areas (2:1)	0.5 - 0.7
Turf meadows	0.1 - 0.4
Forested areas	0.1 - 0.3
Cultivated fields	0.2 - 0.4
Urban Areas	
Flat residential, with about 30 percent of area impervious	0.40
Flat residential, with about 60 percent of area impervious	0.55
Moderately steep residential, with about 50 percent of area impervious	0.65
Moderately steep built up area, with about 70 percent of area impervious	0.80
Flat commercial, with about 90 percent of area impervious	0.80

[1]For flat slopes and permeable soil, use the lower values. For steep slopes and impermeable soil, use the higher values.

Table 12. Manning's Roughness Coefficients for Various Boundaries.	
Rigid Boundary Channels	Manning's n
Very smooth concrete and planed timber	0.011
Smooth concrete	0.012
Ordinary concrete lining	0.013
Wood	0.014
Vitrified clay	0.015
Shot concrete, untroweled, and earth channels in best condition	0.017
Straight unlined earth canals in good condition	0.020
Mountain streams with rocky beds	0.040 -0.050
MINOR STREAMS (top width at flood stage < 30 m)	
Streams on Plain	
1. Clean, straight, full stage, no rifts or deep pools	0.025-0.033
2. Same as above, but more stones and weeds	0.030-0.040
3. Clean, winding, some pools and shoals	0.033-0.045
4. Same as above, but some weeds and stones	0.035-0.050
5. Same as above, lower stages, more ineffective slopes and sections	0.040-0.055
6. Same as 4, but more stones	0.045-0.060
7. Sluggish reaches, weedy, deep pools	0.050-0.080
8. Very weedy reaches, deep pools, or floodways with heavy stand of timber and underbrush	0.075-0.150
Mountain Streams, no Vegetation in Channel, Banks Usually Steep, Trees and Brush Along Banks Submerged at High Stages	
1. Bottom: gavels, cobbles and few boulders	0.030-0.050
2. Bottom: cobbles with large boulders	0.040-0.070
Floodplains	
Pasture, No Brush	
1. Short Grass	0.025-0.035
2. High Grass	0.030-0.050
Cultivated Areas	
1. No Crop	0.020-0.040
2. Mature Row Crops	0.025-0.045
3. Mature Field Crops	0.030-0.050
Brush	
1. Scattered brush, heavy weeds	0.035-0.070
2. Light brush and trees in winter	0.035-0.060
3. Light brush and trees in summer	0.040-0.080
4. Medium to dense brush in winter	0.045-0.110
5. Medium to dense brush in summer	0.070-0.160

Table 12. Manning's Roughness Coefficients for Various Boundaries (continued).	
Rigid Boundary Channels	Manning's n
Trees	
1. Dense willows, summer, straight	0.110-0.200
2. Cleared land with tree stumps, no sprouts	0.030-0.050
3. Same as above, but with heavy growth of sprouts	0.050-0.080
4. Heavy stand of timber, a few down trees, little undergrowth, flood stage below branches	0.080-0.120
5. Same as above, but with flood stage reaching branches	0.100-0.160
MAJOR STREAMS (Topwidth at flood stage > 30 m)	
The n value is less than that for minor streams of similar description, because banks offer less effective resistance. Regular section with no boulders or brush Irregular and rough section	0.025-0.060 0.035-0.100
Alluvial Sand-bed Channels (no vegetation)	
Tranquil flow, Fr < 1 Plane bed Ripples Dunes Washed out dunes or transition Plane bed	0.014-0.020 0.018-0.030 0.020-0.040 0.014-0.025 0.010-0.013
Rapid Flow, Fr > 1 Standing waves Antidunes	0.010-0.015 0.012-0.020
Overland Flow and Sheet Flow	
Smooth asphalt	0.011
Smooth concrete	0.012
Cement rubble surface	0.024
Natural range	0.13
Dense grass	0.24
Bermuda grass	0.41
Light underbrush	0.40
Heavy underbrush	0.80

Table 13. Permissible Shear Stresses for Lining Materials.[26]			
Lining Category	Lining Type	Permissible Unit Shear Stress	
		(lb/ft^2)	(Pa)
Temporary*	Woven Pater Net	0.15	7.2
	Jute Net	0.45	21.6
	Fiberglass Roving:		
	Single	0.60	28.7
	Double	0.85	40.7
	Straw with Net	1.45	69.4
	Curled Wood Mat	1.55	74.2
	Synthetic Mat	2.00	95.8
Vegetative**	Class A	3.70	177.2
	Class B	2.10	100.5
	Class C	1.00	47.9
	Class D	0.60	28.7
	Class E	0.35	16.8
Gravel Riprap	25 mm	0.33	15.8
	50 mm	0.67	32.1
Rock Riprap	150 mm	2.00	95.8
	300	4.00	191.5
Bare Soil	Noncohesive	See "Hydraulic Engineering Circular No. 15"[26]	
	Cohesive		

*Some "temporary" linings become permanent when buried.

**A-E refers to retardance class, with Class A vegetation having high retardance and Class E having low retardance. Typical examples include (HEC-15, Table 1):

Retardance Class	Cover	Condition
A	Weeping lovegrass	Excellent stand, tall (76 cm) (30 in)
B	Weeping lovegrass	Good stand, tall (61 cm) (24 in)
C	Bermuda grass	Good stand, mowed (15 cm) (6 in)
D	Bermuda grass	Good stand, cut (6 cm) (2.5 in)
E	Bermuda grass	Good stand, cut (4 cm) (1.5 in)

Table 14. Manning's Roughness Coefficients.[25]				
		n - value		
Lining Category	Lining Type	Depth Ranges		
		0 - 0.15 m	0.15 0.60 m	>0.60 m
		(0 - .5 ft)	(0.5 - 2.0 ft)	>(2.0 ft)
Rigid	Concrete	0.015	0.013	0.013
	Grouted Riprap	0.040	0.030	0.028
	Stone Masonry	0.042	0.032	0.030
	Soil Cement	0.025	0.022	0.020
	Asphalt	0.018	0.016	0.016
Unlined	Bare Soil	0.023	0.020	0.020
	Rock Cut	0.045	0.035	0.025
Temporary*	Woven Paper Net	0.016	0.015	0.015
	Jute Net	0.028	0.022	0.019
	Fiberglass Roving	0.028	0.021	0.019
	Straw with Net	0.065	0.033	0.025
	Curled Wood Mat	0.066	0.035	0.028
	Synthetic Mat	0.036	0.025	0.021
Gravel Riprap	25 mm (1 in) D_{50}	0.044	0.033	0.030
	50 mm (2 in) D_{50}	0.066	0.041	0.034
Rock Riprap	150 mm (6 in) D_{50}	0.104	0.069	0.035
	300 mm (12 in) D_{50}	--	0.078	0.040

*Some "temporary" linings become permanent when buried.

Note: Values listed are representative values for the respective depth ranges.
Manning's roughness coefficients n vary with the flow depth.

189

Table 15. Manning's n Values for Closed Conduits.	
Description	Manning's n Range
Concrete pipe	0.011-0.013
Corrugated metal pipe or pipe-arch:	
Corrugated Metal Pipes and Boxes, Annular or Helical Pipe (Manning's n varies with barrel size) — 68 by 13 mm (2-2/3 x 1/2 in.) corrugations	0.022-0.027
150 by 25 mm (6 x 1 in.) corrugations	0.022-0.025
125 by 25 mm (5 x 1 in.) corrugations	0.025-0.026
75 by 25 mm (3 x 1 in) corrugations	0.027-0.028
150 by 50 mm (6 x 2 in.) structural plate corrugations	0.033-0.035
230 by 64 mm (9 x 2-1/2 in.) structural plate corrugations	0.033-0.037
Corrugated Metal Pipes Helical Corrugations, Full Circular Flow — 68 by 13 mm (2-2/3 x 1/2 in.) corrugations	0.012-0.024
Spiral Rib Metal Pipe — Smooth walls	0.012-0.013
Vitrified clay pipe	0.012-0.014
Cast-iron pipe, uncoated	0.013
Steel pipe	0.009-0.013
Brick	0.014-0.017
Monolithic concrete:	
1. Wood forms, rough	0.015-0.017
2. Wood forms, smooth	0.012-0.014
3. Steel forms	0.012-0.013
Cemented rubble masonry walls:	
1. Concrete floor and top	0.017-0.022
2. Natural floor	0.019-0.025
Laminated treated wood	0.015-0.017
Vitrified clay liner plates	0.015

NOTE: The values indicated in this table are recommended Manning's n design values. Actual field values for older existing pipelines may vary depending on the effects of abrasion, corrosion, deflection, and joint conditions. Concrete pipe with poor joints and deteriorated walls may have n values of 0.014 to 0.018. Corrugated metal pipe with joint and wall problems may also have higher n values, and in addition, may experience shape changes which could adversely effect the general hydraulic characteristics of the pipeline.

Form Loss Relationships

A.   Access hole losses (HYDRAIN procedure)[10]

Access hole losses are calculated as a coefficient times the outlet velocity head. The coefficient K has been experimentally defined as

$$K = K_o \times C_D \times C_d \times C_Q \times C_p \times C_B$$

where:

K =   Adjusted headloss coefficient
K_o =   Initial headloss coefficient based on relative access hole size
C_D =   Correction factor for pipe diameter
C_d =   Correction factor for flow depth
C_Q =   Correction factor for relative flow
C_p =   Correction factor for plunging flow
C_B =   Correction factor for benching

The initial headloss coefficient K_o, is estimated as a function of the relative access hole size and angle between the inflow and outflow pipes:

$$K_o = 0.1 \times \left[\frac{b}{D_o}\right] \times [1 - \sin\theta] + 1.4 \times \left[\frac{b}{D_o}\right]^{0.15} \times \sin\theta$$

where:

K_o =   Initial headloss coefficient based on relative access hole size
θ =   Angle between the inflow and outflow pipes (see figure 55)
b =   Access hole diameter
D_o =   Outlet pipe diameter

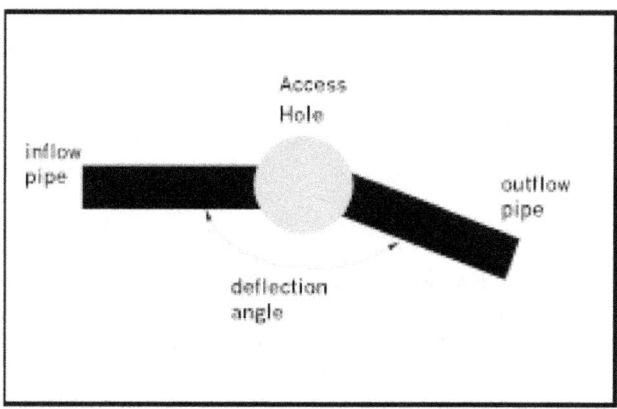

Figure 55.   Angle of deflection.

It has been shown that there are only slight differences in headloss coefficient between round and square access holes. Therefore, access hole shape can be ignored when estimating headlosses for design purposes.

191

The correction factor for pipe diameter, C_D, was determined to be:

$$C_D = \left[\frac{D_o}{D_i} \right]^3$$

where:

C_D	=	Correction factor for variation in pipe diameter
D_i	=	Incoming pipe diameter
D_o	=	Outgoing pipe diameter

A change in headloss due to differences in pipe diameter was found to only be significant in pressure flow situations when the depth in the access hole to outlet pipe diameter ratio, d/D_o, is greater than 3.2. Therefore, it is only applied in such cases.

The correction factor for flow depth, C_d, is calculated by the following:

$$C_d = 0.5 \times \left[\frac{d}{D_o} \right]^{\frac{3}{5}}$$

where:

C_d	=	Correction factor for flow depth
d	=	Water depth in access hole above outlet pipe invert
D_o	=	Outlet pipe diameter

This correction factor was found to be significant only in cases of free surface flow or low pressures, when d/D_o ratio is less than 3.2, and is only applied in such cases. Water depth in the access hole is approximated as the level of the hydraulic gradeline at the upstream end of the outlet pipe.

The correction factor for relative flow, C_Q, is computed by:

$$C_Q = (1 - 2 \times \sin \theta) \times \left[1 - \frac{Q_i}{Q_o} \right]^{\frac{3}{4}} + 1$$

where:

C_Q	=	Correction factor for relative flow
θ	=	The angle between the original inflow and outflow pipes
Q_i	=	Flow in the original inflow pipe
Q_o	=	Flow in the outlet pipe

C_Q is a function of the angle of the incoming flow as well as the percentage of flow coming in through the pipe of interest versus other incoming pipes. To illustrate this effect, consider the access hole shown in figure 56 and assume that Q_1 = 3 m³/s (105.94 ft³/s), Q_2 = 1 m³/s (35.3 ft³/s), and Q_3 = 4 m³/s (141.26 ft³/s). Solving for the relative flow correction factor in going from the outlet pipe (number 3) to one of the inflow pipes (number 2):

SI	English
$C_{Q_{3 \to 2}} = [1 - 2 \times \sin(90°)] \times \left[1 - \dfrac{1}{4} \right]^{3/4} + 1 = 0.19$	$C_{Q_{3 \to 2}} = [1 - 2 \times \sin(90°)] \times \left[1 - \dfrac{35.3}{141.26} \right]^{3/4} + 1 = 0.19$

For a second example, consider the following flow regime: $Q_1 = 1$ m³/s, $Q_2 = 3$ m³/s, $Q_3 = 4$ m³/s. Calculating C_Q for this case:

SI	English
$C_{Q_{3 \to 2}} = [1 - 2 \times \sin(90°)] \times \left[1 - \dfrac{3}{4} \right]^{3/4} + 1 = 0.65$	$C_{Q_{3 \to 2}} = [1 - 2 \times \sin(90°)] \times \left[1 - \dfrac{105.94}{141.26} \right]^{3/4} + 1 = 0.65$

In both of these cases, the flow coming in through pipe number 2 has to make a 90-degree bend before it can go out pipe number 3. In case 1, the larger flow traveling straight through the access hole, from pipe number 1 to pipe number 3, assists the flow from pipe number 2 in making this bend. In case 2, a majority of the flow is coming in through pipe number 2. There is less assistance from the straight through flow in directing the flow from pipe number 2 into pipe number 3. As a result, the correction factor for relative flow in case 1 (0.19) was much smaller than the correction factor for case 2 (0.65).

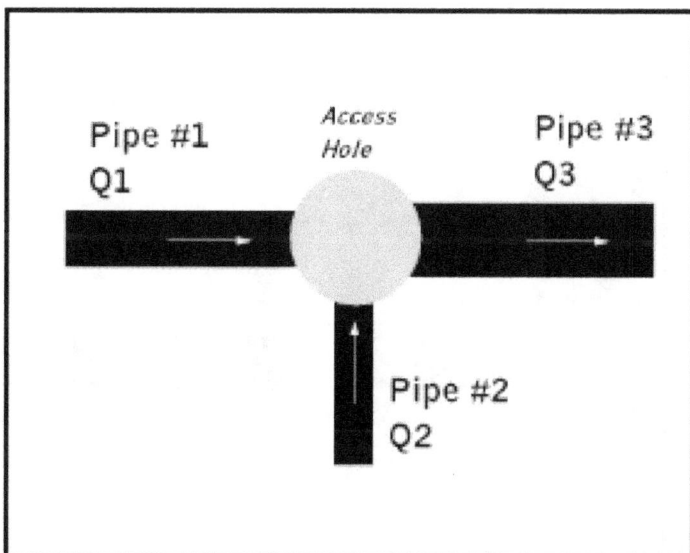

Figure 56. Example of relative flow effect.

The correction factor for plunging flow, C_p, is calculated by the following:

$$C_p = 1 + 0.2 \times \left[\frac{h}{D_o}\right] \times \left[\frac{h-d}{D_o}\right]$$

where:

C_p	=	Correction for plunging flow
h	=	Vertical distance of plunging flow from the invert of the plunging pipe to the center of the outlet pipe
D_o	=	Outlet pipe diameter
d	=	Water depth in the access hole

This correction factor corresponds to the effect of another inflow pipe, plunging into the access hole, on the inflow pipe for which the headloss is being calculated. Using the notations the above figure, for example, C_p is calculated for pipe number 2 when pipe number 1 discharges plunging flow. The plunging flow that results from flow entering through the inlet into the access hole is considered in the same manner. The correction factor is only applied when h is greater than d.

The final correction factor multiplied by the initial headloss coefficient K_o to get the adjusted headloss coefficient K is the correction for benching in the access hole, C_B. Benching tends to direct flows through the access hole, resulting in reductions in headloss. Figure 57 shows the types of benching considered in HYDRA. The benching correction factors employed by HYDRA are shown in table 16. For flow depths between the submerged and unsubmerged conditions, a linear interpolation is performed.

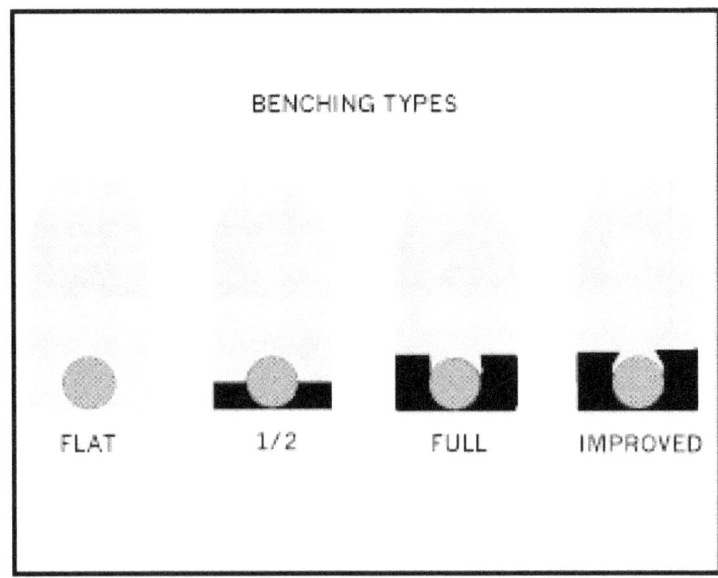

Figure 57. Types of Benching.

194

Table 16. Correction Factors, C_B, for Benching.		
	Correction Factors C_B	
Bench Type	Submerged*	Unsubmerged**
Flat floor	1.00	1.00
Benched one-half of pipe diameter	0.95	0.15
Benched one pipe diameter	0.75	0.07
Improved	0.40	0.02
*Pressure flow, d/Do > 3.2		
**Free surface flow, d/Do <1.0		

B.   Transition Losses

$$h_t = K_t \left| \frac{V_u^2 - V_d^2}{2g} \right|$$

where:

$\left| \dfrac{V_u^2 - V_d^2}{2g} \right|$ = Absolute value of the velocity head differential from upstream to downstream.

K_t = 0.1 for increasing velocity and 0.2 for decreasing velocity.

For **pressure flow** through transitions, the following equations are applicable:

Expansion transition:

$$h_{et} = K_e \left[\frac{(V_u - V_d)^2}{2g} \right]$$

where:

K_e =   1.0 for a sudden expansion
K_e =   0.2 for well designed transitions

C.   Bend Losses

A measurable bend loss occurs when a conduit run changes direction by more than about 15°. The estimate for this loss is expressed as:

$$h_b = K_b \frac{V^2}{2g}$$

where:

V =   Velocity of flow in conduit

K_b = $0.25 \sqrt{\dfrac{\phi}{90}}$

ϕ =   Central angle of bend (degrees)

APPENDIX C

Glossary

(page intentionally left blank)

APPENDIX C

Glossary

A listing of terms related to highways and river environment, streambank protection, and river mechanics is provided below:

Acceleration - Acceleration is the time rate of change in magnitude or direction of the velocity vector. Units are meters per second per second (m/s^2) or feet per second per second (ft/s^2). It is a vector quantity. Acceleration has components both tangential and normal to the streamline, the tangential component embodying the change in magnitude of the velocity, and the normal component reflecting a change in direction.

Articulated Concrete Mattress - Rigid concrete slabs, which can move as scour occurs without separating, usually hinged together with corrosion-resistant wire fasteners; primarily placed for lower bank protection.

Average Velocity - Velocity at a given cross section determined by dividing discharge by cross-sectional area.

Backwater - The increase in water surface elevation relative to the elevation occurring under natural channel and floodplain conditions, induced upstream from a bridge or other structure that obstructs or constricts a channel.

Backwater Area - The low-lying lands adjacent to a stream that may become flooded due to backwater effects.

Bedform - A relief feature on the bed of a stream, such as dunes, plane bed or antidunes. Also called bed configuration.

Bituminous Mattress - An impermeable rock-, mesh-, or metal-reinforced layer of asphalt or other bituminous material placed on a streambank to prevent erosion.

Cellular-block Mattress - Regularly cavitated interconnected concrete blocks placed directly on a streambank or filter to prevent erosion. The cavities can permit bank drainage and the growth of either volunteer or planted vegetation when synthetic filter fabric is not used between the mattress and bank.

Channel - The bed and banks that confine the flow.

Choking (of flow) - Severe backwater effect resulting from excessive constriction.

Closed-conduit Flow - Flow in a pipe, culvert, etc. where there is a solid boundary on all four sides. Examples are pipes, culverts, and box culverts. They may be flowing full-pressure flow, or partly full open-channel flow.

Concrete Paving - Plain or reinforced concrete slabs poured or placed on the surface to be protected.

Constriction - A control section, such as a bridge crossing, channel reach or dam, with limited flow capacity in which the discharge is related to the upstream water surface elevation; a constriction may be either natural or artificial.

199

Cross Section - A diagram or drawing cut across a channel that illustrates the banks, bed, and water surface.

Daily Discharge - The discharge of water or sediment averaged over one day.

Design Discharge - Flow a drainage facility is expected to accommodate without exceeding adopted design criteria.

Design High-water Level - Water level that a drainage facility is designed to accommodate without exceeding adopted design criteria.

Discharge - Time rate of the movement of a quantity of water or sediment passing a given cross section of a stream or river.

Discharge - The quantity of water moving past a given plane (cross section) in a given unit of time. Units are cubic meters per second (m^3/s) or cubic feet per second (ft^3/s). The plane or cross section must be perpendicular to the velocity vector.

Drainage Basin - An area confined by drainage divides, often having only one outlet for discharge.

Energy Grade Line - An inclined line representing the total energy of a stream flowing from a higher to a lower elevation. For open-channel flow, the energy grade line is located a distance of $V^2/2g$ above the water surface (V = velocity and g = acceleration due to gravity).

Erosion - Displacement and movement of sediment or soil particles due to the movement of water.

Erosion Control Matting - Fibrous matting (e.g. jute, paper, etc.) placed or sprayed on a streambank for the purpose of preventing erosion or providing temporary stabilization until vegetation is established.

Filter - Layer of fabric, sand, gravel, or graded rock placed, or developed naturally, where suitable in-place materials exist, between the bank revetment and soil for one or more of three purposes: to prevent the soil from moving through the revetment by piping, extrusion, or erosion; to prevent the revetment from sinking into the soil; and to permit natural seepage from the streambank, thus preventing buildup of excessive hydrostatic pressure.

Flood-frequency Curve - A graph indicating the probability that the annual flood discharge will exceed a given magnitude, or the recurrence interval corresponding to a given magnitude.

Floodplain - Alluvial lowland bordering a stream that is subject to inundation by floods.

Flow-duration Curve - A graph indicating the percentage of time a given discharge is exceeded.

Freeboard - The vertical distance above a design stage that is allowed for waves, surges, drift, and other contingencies.

Froude Number - A dimensionless number (expressed as $V/\sqrt{gy}$) that represents the ratio of inertial to gravitational forces. High Froude numbers can be indicative of high flow velocity and scour potential.

Gabion - A basket or compartmented rectangular container made of steel wire mesh. When filled with cobbles or other rock of suitable size, the gabion becomes a flexible and permeable block with which flow-control structures can be built.

Grout - A fluid mixture of cement and water or of cement, sand, and water used to fill joints and voids.

Historical Flood - The largest known flood event at a given site.

Hydraulics - The applied science concerned with the behavior and flow of liquids, especially in pipes, channels, structures, and the ground. The engineering application of the mechanical properties of fluids (water) in motion. The determination of the forces and energy of water in motion or at rest.

Hydraulic Problem - An effect of stream flow, tidal flow, or wave action on a crossing such that traffic is immediately or potentially disrupted, or another highway detrimental effect is caused.

Hydrograph - The graph of stage or discharge against time.

Hydrology - The determination of where surface or groundwater will occur, in what quantity, and with what frequency.

Hydraulic Radius - The hydraulic radius is a length term used in many of the hydraulic equations that is determined by dividing the flow area by the length of the cross section in contact with the water (wetted perimeter). The hydraulic radius is in many of the equations to help take into account the effects of the shape of the cross section on the flow. The hydraulic radius for a circular pipe flowing full is equal to the diameter of the pipe divided by four (D/4).

Instantaneous Discharge - A discharge at a given moment.

Laminar Flow - In laminar flow, the mixing of the fluid and momentum transfer is by molecular activity.

Local Acceleration - Local acceleration is the change in velocity (either or both magnitude and direction) with time at a given point or cross section.

Nonuniform Flow - In nonuniform flow, the velocity of flow changes in magnitude or direction or both with distance. The convective acceleration components are different from zero. Examples are flow around a bend or flow in expansions or contractions.

One-dimensional Flow - A method of analysis where changes in the flow variables (velocity, depth, etc.) occur primarily in the longitudinal direction. Changes of flow variables in the other two dimensions are assumed to be small and are neglected.

Open-channel Flow - Open-channel flow is flow with a free surface. Closed-conduit flow or flow in culverts is open-channel flow if they are not flowing full and there is a free surface.

Overbank Flow - Water movement over top bank either due to a rising stream stage or to inland surface water runoff.

Pressure Flow - Flow in a closed conduit or culvert that is flowing full with water in contact with the total enclosed boundary.

Reach - A segment of stream length that is arbitrarily selected for purpose of study.

Recurrence Interval (R.I.); Return Period; Exceedence Interval - The reciprocal of the annual probability of exceedence of a hydrologic event.

Reynolds Number - The Reynolds Number is the dimensionless ratio of the inertial forces to the viscous forces. It is defined as $(Re = \rho VL/\mu)$, where ρ and μ are the density and dynamic viscosity of the fluid, V is the fluid velocity, and L is a characteristic dimension, usually the depth (or the hydraulic radius) in open-channel flow.

Resistance to Flow - The effect of the boundaries on the flow of water. It is measured by the Manning's n, Chezy C, or Darcy-Weisbach f.

Roughness Coefficient - Numerical measure of the frictional resistance to flow in a channel such as the Manning's coefficient.

Runoff - That part of precipitation which appears in surface streams of either perennial or intermittent form.

Shear Stress, Tractive Force - The force or drag on the channel boundaries caused by the flowing water. For uniform flow, shear stress is equal to the unit weight of water times the hydraulic radius times the slope. Usually expressed as force per unit area.

Slope - Fall per unit length of the channel bottom, water surface or energy grade line.

Soil-cement - A designed mixture of soil and portland cement compacted at a proper water content to form a veneer or structure that can prevent streambank erosion.

Steady Flow - In steady flow, the velocity at a point or cross section does not change with time. The local acceleration is zero.

Stone Riprap - Natural cobbles, boulders, or rock dumped or placed on a streambank or filter as protection against erosion.

Stream - A body of water that may range in size from a large river to a small rill flowing in a channel. By extension, the term is sometimes applied to a natural channel or drainage course formed by flowing water whether it is occupied by water or not.

Streamline - An imaginary line within the flow which is everywhere tangent to the velocity vector.

Subcritical Flow - In open-channel flow, the free surface configuration, in response to changes in channel geometry depends on the Froude Number $(Fr = V\sqrt{gL})$ which is the ratio of inertial forces to gravitational forces. The Froude Number is also the ratio of the flow velocity v to the celerity $(c = \sqrt{gL})$ of a small gravity wave in the flow. When $Fr < 1$, the flow is subcritical (or tranquil), and surface waves propagate upstream as well as downstream. **The boundary condition that controls the subcritical flow depth is always located at the downstream end of the subcritical reach.**

<u>Supercritical Flow</u> - When Fr > 1, the flow is supercritical (or rapid) and surface disturbances can propagate only in the downstream direction. **The control section for supercritical flow is always at the upstream end of the supercritical flow reach.** When Fr = 1.0, the flow is critical and surface disturbances remain stationary in the flow.

<u>Thalweg</u> - The line extending down a channel that follows the lowest elevation of the bed.

<u>Three-dimensional Flow</u> - A method of analysis where the flow variables change in all three dimensions, along, across, and in the vertical.

<u>Two-dimensional Flow</u> - A method of analysis where the accelerations occur in two directions (along and across the flow).

<u>Tractive Force</u> - The drag on a streambank caused by passing water which tends to pull soil particles along with the streamflow.

<u>Turbulence</u> - Motion of fluids in which local velocities and pressures fluctuate irregularly in a random manner as opposed to laminar flow where all particles of the fluid move in distinct and separate lines.

<u>Turbulent Flow</u> - In turbulent flow the mixing of the fluid and momentum transfer is related to random velocity fluctuations. The flow is laminar or turbulent depending on the value of the Reynolds number (Re). In laminar flow, viscous forces are dominant and Re is relatively small. In turbulent flow, Re is large; that is, inertial forces are very much greater than viscous forces. Turbulent flows are predominant in nature. Laminar flow occurs very infrequently in open-channel flow.

<u>Uniform Flow</u> - In uniform flow the velocity of the flow does not change with distance. The convective acceleration is zero. Examples are flow in a straight pipe of uniform cross section flowing full or flow in a straight open channel with constant slope and all cross sections of identical form, roughness and area, resulting in a constant mean velocity. Uniform flow conditions are rarely attained in open channels, but the error in assuming uniform flow in a channel of fairly constant slope and cross section is small in comparison to the error in determining the design discharge.

<u>Unit Discharge</u> - Discharge per unit width (may be average over a cross section, or local at a point).

<u>Unsteady Flow</u> - In unsteady flow, the velocity at a point or cross section varies with time. The local acceleration is not zero. A flood hydrograph where the discharge in a stream changes with time is an example of unsteady flow. Unsteady flow is difficult to analyze unless the time changes are small.

<u>Velocity, Cross-sectional Average</u> - Discharge divided by cross-sectional area of flow.

(page intentionally left blank)